One Sky Many Calendars
25 Ideas That Explain How We Measure Time

Rahul Agrawal

For **Arjun** and **Navya** who reminded me to look up the sky again...

Introduction

Early evening in Melbourne. The sky is still blue. A thin crescent moon is visible. Navya (6) is on her back in the grass, looking straight up the sky. Arjun (12) is standing at the back door studying the wall calendar inside, puzzled. Dad is sitting on a low wooden chair and watching both of them.

Navya *(pointing at the Moon)*: Does the Moon look the same every night?

Arjun: No… sometimes it's full. Sometimes it's just a thin curve.

Navya: Why does it keep changing?

Arjun *(thinking)*: I think months have something to do with the Moon.

Dad *(smiling)*: That's a very good question. One that people have been asking for thousands of years.

Arjun *(glancing at the calendar)*: Dad… Diwali keeps changing dates. But our calendar is fixed and does not change. Why does this happen?

Dad: Because they don't follow the same thing.

Navya: The Moon and the calendar?

Dad: Yes.

Arjun: So, which one is right?

Dad *(smiling)*: Let's find out.

Somewhere in the world right now, a child is asking why their festival falls on a different date this year. Maybe it is Diwali. Maybe it is Eid. Perhaps it is Chinese New Year or Easter or Rosh Hashanah. The question is the same everywhere. Why does it keep moving? The honest answer is that most adults do not know either.

These questions have been asked by cultures around the world for as long as people have looked up at the sky. They are not simple questions. They sit at the intersection of astronomy, history, religion and mathematics. Every civilisation that has ever tried to build a calendar has had to solve the same puzzle. How do you count time using a Moon and a Sun that refuse to cooperate with each other? The different answers to that puzzle are what this book is about.

The full answer involves the Moon, the Sun, the Earth's axial tilt and a number the sky refuses to make whole. It involves a Roman emperor's pride, a Jesuit mathematician working for a decade in candlelight, ten days erased from October 1582 by papal decree and the independent discovery of the same mathematical cycle by astronomers in Babylon, India, Greece and China who never spoke to each other. It involves festivals that have been observed for three thousand years and a calendar that started in Egypt and ended up on your wall.

This book follows a family living in Melbourne, Arjun (12), his sister Navya (6) and their parents. Their neighbours Oliver and Mei appear from time to time, as neighbours do. The conversations happen in a backyard over several evenings, as the Moon moves through its phases above them. Wherever you are reading this, whatever calendar your family lives by, the sky above you is the same sky they are sitting under.

That sky was watched by Egyptian priests waiting for Sirius to rise before dawn, by Aboriginal Australians reading the dark shape of an emu written across the Milky Way, by Babylonian astronomers pressing their observations into clay, by Vedic Indians along the Saraswati River tracking the rhythms of stars and seasons, by ancient Chinese skywatchers mapping celestial patterns into ordered systems and by Hebrew scholars marking sacred time through lunar cycles and scripture. It has been watched by every family that has ever looked up and wondered.

The calendars different civilisations built to describe this sky are portraits of those civilisations. It is their values, their priorities and their negotiations between nature and culture. This book is an introduction to those portraits.

It begins, as all good things do, with a child asking a question.

The Calendars the World Lives By

Most of the world uses two calendars simultaneously and anyone barely notices.

The first is the Gregorian calendar. The one on official documents, school planners and most of the world's administrative systems. Pope Gregory XIII introduced it in 1582 to fix errors in the older Roman calendar. Today nearly every country uses it for business, government and international coordination. It follows the Sun. One year is one orbit of the Earth around the Sun. 365 days with a leap day added every four years (with some exceptions). It is convenient, globally standardised and almost completely disconnected from the Moon.

The second calendar is the one, billions of people actually feel. It governs when they fast, when they feast, when they plant crops and when they celebrate. It tracks the Moon, or the Moon and the Sun together. It is often older than the Gregorian calendar by thousands of years. And it is almost certainly *more astronomically sophisticated*.

Between them, the four great non-Gregorian calendar traditions cover the majority of the human population.

The Indian Panchang - Used by more than a billion Hindus, Jains and Sikhs. The panchang is the most detailed calendar in common daily use anywhere in the world. The word Panchang itself means five parts in Sanskrit, referring to the five streams of astronomical information it tracks simultaneously, the lunar day, the day of the week, the Moon's position among 27 lunar mansions and two further sacred measures. It is a lunisolar calendar. The lunar months correct periodically by an extra month called Adhik Maas to keep festivals aligned with the solar seasons. The nakshatra system at its core is estimated to be 3,000 to 4,000 years old. Most major Hindu festival from Diwali to Holi to Navratri is calculated by the panchang.

The Chinese Lunisolar Calendar - One of the longest continuously maintained calendar systems in history, with records going back more than 4,000 years. Like the Indian panchang, it uses lunar months but inserts a leap month when necessary, governed by 24 solar terms that divide the year by the Sun's position. Chinese New Year, the most widely celebrated festival on Earth by total participants, falls on the second new moon after the winter solstice. The same underlying system governs Vietnamese Tet, Korean Seollal and the lunar new year celebrations of Chinese communities worldwide. Around 1.5 billion people orient their festival year by this calendar.

The Hebrew Calendar - Used by approximately 15 million Jewish people worldwide, the Hebrew calendar has maintained extraordinary consistency for around 1,600 years. It is a lunisolar system built on a 19-year mathematical cycle. A near-perfect coincidence of 235 lunar months with 19 solar years, which allows a leap month to be inserted in the right years to keep Passover reliably in spring. The Hebrew calendar currently runs in the 5780s, counting from the traditional date of creation. Though its community is smaller than the others, its lunar system and its role in setting Passover helped shape how Easter is calculated for 2.4 billion Christians.

The Islamic Calendar - Used by approximately 1.8 billion Muslims for religious observance, the Islamic calendar began in 622 CE, the year the Prophet Muhammad migrated from Mecca to Medina. It follows the Moon alone. It has twelve lunar months, each beginning with the sighting of the crescent moon, totalling 354 days per year. No correction is made for the solar year. As a result, Islamic festivals including Ramadan and Eid move through every season over a 32-year cycle. This is deliberate. No season should be permanently favoured over another for the performance of religious duty.

Others - Dozens of other systems remain in active use. The Persian Solar Hijri calendar, reformed in 1079 CE with contributions from the mathematician and poet Omar Khayyam, is the official civil calendar of Iran and Afghanistan. The Ethiopian calendar runs roughly seven years behind the Gregorian and has 13 months. The Coptic, Julian and Tibetan calendars each serve tens of millions. Aboriginal Australian communities across the continent maintain seasonal sky calendars of extraordinary precision that predate any written record by thousands of years.

All of these calendars describe the same sky.
This book is about why they disagree and why every one of them is right.

The Festivals the World Lives By

You may know some of these festivals well. You may be celebrating one yourself this month. Others may be entirely new to you. All of them are worth knowing, because all of them are answers to the same question. How do you mark the passage of time in a way that connects you to the sky, the seasons and the people around you?

Every festival in this book marks a specific astronomical moment. A new moon, a full moon, a solstice, an equinox or the rising of a star. This page introduces the main ones briefly, so as when they appear in the conversation, they arrive as something familiar rather than something foreign.

Solar Festivals
These fall on the same Gregorian date every year because they follow the Sun.

Christmas (December 25th) Celebrated by approximately 2.4 billion Christians worldwide to mark the birth of Jesus. It's fixed to the solar calendar. It occurs always in northern hemisphere winter and always in southern hemisphere summer.
Nowruz (around March 20th to 21st) The Persian New Year, celebrated by over 300 million people across Iran, Afghanistan, Central Asia and diaspora communities worldwide. It falls precisely on the vernal equinox. The moment day and night are equal and spring begins. It is one of the oldest continuously observed celebrations on Earth, predating Islam by at least 2,000 years.

Lunisolar Festivals
These follow the Moon but stay anchored to the same season every year through periodic calendar correction.

Diwali (new moon of the lunar month Kartik - October or November) The festival of lights, celebrated by over a billion Hindus, Jains and Sikhs worldwide. It falls on the darkest night of the lunar month of Kartik. It is celebrated always in autumn in the northern hemisphere and always in spring in the southern hemisphere, symbolising the triumph of light over darkness and the return of Rama to Ayodhya.
Chinese New Year (second new moon after the winter solstice - January or February) The most widely celebrated festival on Earth by total participants, observed by approximately 1.5 billion people across China, Vietnam, Korea and Chinese communities worldwide. It falls always between January 21st and February 20th. Each year is associated with one of twelve zodiac animals in a repeating cycle.

Passover (full moon of the Hebrew month Nisan - March or April) The Jewish festival of liberation, marking the exodus of the Israelites from Egypt. It falls on the full moon of Nisan, which the Hebrew calendar keeps in spring. The Passover Seder begins on the evening the full moon rises.

Rosh Hashanah (new moon of the Hebrew month Tishrei: September or October) The Jewish New Year, marking the beginning of the High Holy Days. It begins on the first day of Tishrei, set by the new moon, reflecting the Hebrew calendar's balance of lunar months with the solar year and keeping it in early autumn.

Easter (first Sunday after the first full moon after Spring Equinox) The most important festival in the Christian calendar, marking the resurrection of Jesus. Its date is set by a rule combining the solar year, the lunar cycle and the seven-day week. It falls anywhere between March 22nd and April 25th.

Lunar Festivals

These follow the Moon alone and move through all four seasons over a 32-year cycle.

Ramadan and Eid (ninth and tenth months of the Islamic lunar calendar - move through all seasons) Ramadan is the Islamic month of fasting, observed by approximately 1.8 billion Muslims worldwide. From before dawn to after sunset each day, Muslims abstain from food and drink. Because the Islamic calendar is purely lunar and makes no seasonal correction, Ramadan moves backward by approximately eleven days every year, cycling through every season over roughly thirty-two years. Eid al-Fitr, the festival of breaking the fast, marks its end. Its date is confirmed by the sighting of the crescent moon.

Every festival above is tied to a specific moment in the sky - a new moon, a full moon, a solstice or an equinox. The rest of this book is the story of how different civilisations built calendars to track those moments and why, despite watching the same sky, they arrived at such different answers.

A Timeline of Time

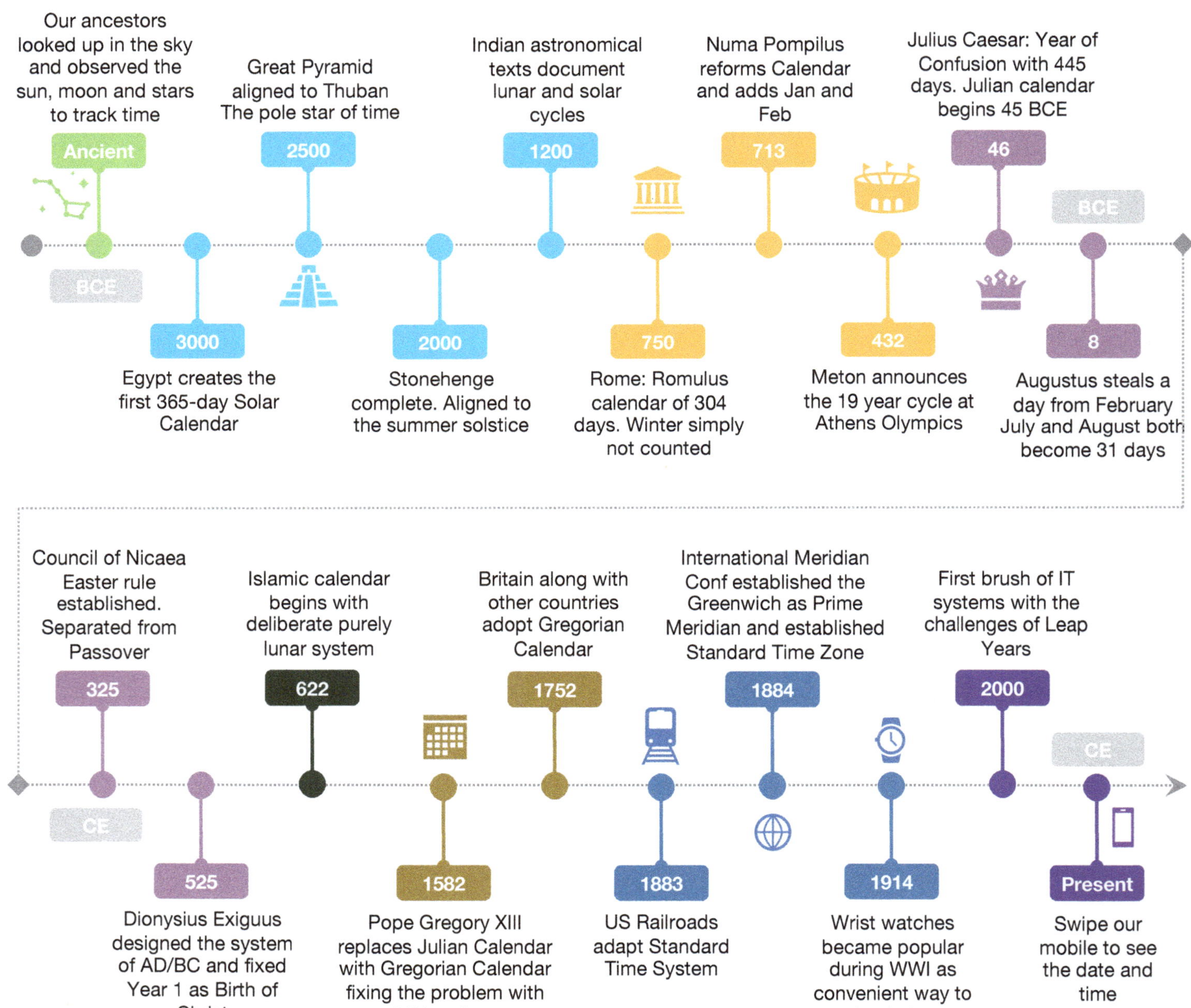

ARC 1
THE CONFUSION

01 Why Do Festival Dates Keep Changing?

After dinner. Navya sits at the table colouring a picture of Diwali fireworks. Arjun stands at the fridge, studying the calendar, puzzled.

Arjun: Dad, why is Diwali on a different date every year? This year it's in October. Last year it was in November.

Dad: Because Diwali follows the moon. It falls on the new moon of Kartik, the darkest night of that lunar month.

Arjun: But the moon cycles every thirty days. So why doesn't Diwali just move by thirty days each year?

Dad pauses. This is a better question than he expected.

Dad: Because the moon's cycle and the year's cycle don't line up neatly. The moon doesn't care about years. It just keeps going.

Navya: *(not looking up from her colouring)* The moon is rude.

Dad smiles.

Navya: At school we made lanterns. Red and gold ones. For Chinese New Year. My teacher said that moves too.

Arjun: So, it's not just Diwali. Why does Chinese New Year move but Christmas doesn't?

Dad: That is exactly the right question. And once you see the answer you'll see it in every calendar in the world.

Dad leans back, thinking. He will have to start from the beginning. This is going to be fun.

Some festivals fall on the same date every year. Christmas is always December 25th. Nowruz, the Persian New Year, is always on March 20 or 21. But Diwali, Chinese New Year and Easter appear to wander, showing up at noticeably different points in the Gregorian year each time.

The difference is which astronomical cycle each festival follows. The Gregorian calendar, the one on the fridge and in most official documents worldwide, follows the Sun. One year is one complete orbit of Earth around the Sun, approximately 365 days. Every Gregorian date sits at the same point in that orbit every year, which means it always falls in the same season. December 25th is always in northern winter. March 20th is always near the spring equinox. The calendar and the seasons are locked together permanently.

Most of the world's oldest and most widely celebrated festivals were designed around the Moon instead. The Moon completes its cycle from new moon through full moon and back to new moon in approximately 29.5 days. This rhythm is visible from anywhere on Earth. You can watch the moon on any clear night without any instrument at all. For ancient peoples watching the sky, it was the most reliable and obvious way to count time between a single day and a whole year.

Diwali falls on the new moon of the lunar month Kartik, the darkest night of that cycle. Chinese New Year falls on the second new moon after the winter solstice. Holi falls on the full moon of the lunar month Phalguna. Each of these moments is fixed precisely within the lunar cycle. The festivals arrive exactly when they always arrive. It is the Gregorian calendar that slides past them each year at a slightly different angle, because the Moon's 29.5-day cycle and the Earth's 365-day orbit share no common factor. One solar year contains roughly 12.37 lunar months, not twelve and not thirteen.

That fraction is why Diwali lands in October one year and November the next, without either date being wrong. It is why Chinese New Year occurs in late January or early February every year and will continue to do so for as long as the Chinese calendar runs. Both calendars are describing the same sky. They are simply counting it by different clocks.

Word Origin: The word "month" and the word "moon" share the same ancient root in almost every language on Earth. The Proto-Indo-European root me(n)ses- meant both at once. Sanskrit has masah (moon and month), Old English mona (moon) and monath (month) are the same word grown slightly apart, Lithuanian mėnesis still means both simultaneously, and Latin mensis (month) and Greek mene (moon) both descend from it. The Gregorian calendar kept the months but quietly forgot the Moon.

02 Why Some Festivals Stay and Others Drift?

A little later. The dishes are cleared after dinner. Dad has drawn a rough diagram on the back of an envelope, a long line for the solar year and twelve shorter marks for the lunar months. The marks fall short of the end of the line by a visible gap. Arjun studies it. His friends Oliver and Mei, who stayed for dinner, lean in with interest.

Oliver: So, Christmas doesn't move because it follows the Sun. And Diwali moves because it follows the Moon. But you said Chinese New Year follows the Moon too and it doesn't move much. Why not?

Dad: *(tapping the gap at the end of the line)* Because the Chinese calendar fixes this gap. Every year, twelve lunar months fall about eleven days short of the solar year. So, the calendar adds an extra month every few years to catch up. That keeps it in the same part of the year.

Mei: I know that. Mum mentioned it once. Something about a repeated month this year.

Dad: Rùn Yuè. Yes. That's how the Chinese calendar fixes the gap.

Arjun: And is that why Diwali stays in the same season?

Dad: Yes. The Indian calendar also inserts an extra month every two or three years to reset the drift. Ramadan doesn't.

Oliver: So, Ramadan just keeps going around?

Dad: All the way around. Thirty-two years for a full circuit. And that's deliberate.

Oliver: (genuinely puzzled) Why would you want that?

Dad: Because not every calendar is trying to stay in the same season.

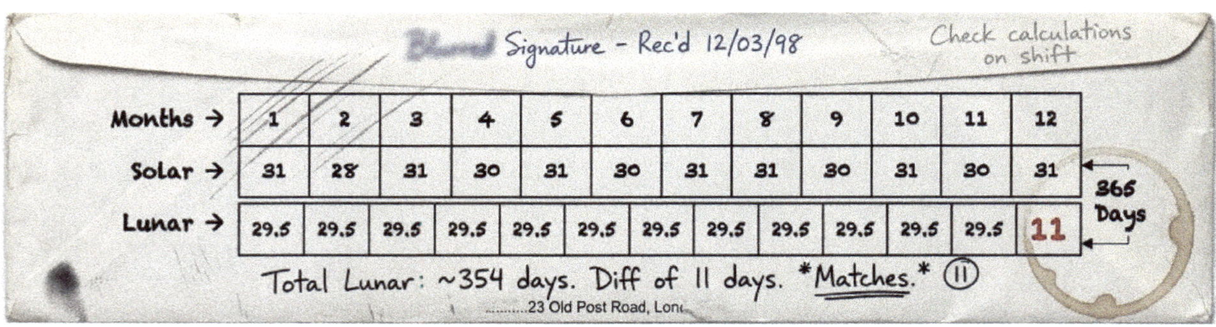

Signature – Rec'd 12/03/98

Check calculations on shift

Months →	1	2	3	4	5	6	7	8	9	10	11	12	
Solar →	31	28	31	30	31	30	31	31	30	31	30	31	365 Days
Lunar →	29.5	29.5	29.5	29.5	29.5	29.5	29.5	29.5	29.5	29.5	29.5	29.5	11

Total Lunar: ~354 days. Diff of 11 days. *Matches.* ⑪

...........23 Old Post Road, Lon...

Not all lunar festivals drift equally. Diwali stays in autumn. Chinese New Year stays in late January or early February. Rosh Hashanah stays in September or October. But Eid al-Fitr appears at a completely different point in the Gregorian calendar every year, cycling through all twelve months over roughly thirty-two years. The difference between these festivals reveals the most important architectural choice any calendar-maker faces - how, and whether, to reconcile the Moon with the Sun.

The Moon's synodic cycle, from one new moon to the next, takes 29.53 days. Twelve of these months total 354.37 days. Approximately eleven days fewer than the 365.24-day solar year. A calendar that follows the Moon alone will lose eleven days against the solar year annually. After three years it has lost about a month. After seventeen years it has drifted by six months. After thirty-two years it has cycled all the way around and returned to its starting point in the Gregorian calendar.

Calendars that keep their festivals in the same season do so by inserting a correction, an extra month added periodically to absorb the eleven-day annual gap. This is called *intercalation*. The Hebrew, Hindu and Chinese lunisolar calendars all use it, though by different methods and with different precision. The result is that Chinese New Year always falls between January 21st and February 20th, and Diwali always falls in October or November. The correction keeps the Moon in dialogue with the Sun.

The Islamic calendar makes a different choice. It follows the Moon alone and makes no seasonal correction. This was not an oversight. It was a deliberate theological decision, reflected in the Quran. The Islamic year is 354 days. Ramadan migrates through all four seasons on a roughly thirty-two-year cycle. A Muslim in a cold country may fast through short winter days in one decade and long summer days in the next. The calendar is honest about the Moon. It simply does not ask the Moon to pretend it agrees with the Sun.

Both approaches are internally consistent. The difference is what each culture decided the calendar was for.

The Number: Eleven. A purely lunar year is eleven days shorter than a solar year. That small gap adds up quickly. Over time, it pushes festivals like Eid al-Fitr steadily backward through the calendar, until they cycle across nearly every month within a single lifetime.

03 The Calendar That Hides Its Complexity

After dinner, Dad brings the family almanac, the Panchang from the shelf and places it beside the Gregorian wall calendar he has taken down. One is dense with columns of small print in multiple colours. The other is a clean grid of numbered squares. Arjun looks between them.

Arjun: Why does one look so much more complicated than the other?

Dad: Because one is showing you what it's doing and the other isn't.

Navya: *(picking up the almanac and turning it upside down)* This one has too many words.

Dad: That one is showing you what it is tracking. The lunar day, the day of the week, the Moon's position in the sky and more. It shows all of them because they all matter.

Arjun: And the Gregorian calendar?

Dad: It is tracking less and hiding more.

Arjun: But it still works.

Dad: It works, yes. But tell me something. Why does February have twenty-eight days?

Arjun pauses.

Dad: Why do some months have thirty days and others thirty-one? Why are July and August both long?

Arjun *(frowning slightly)* I… don't know.

Dad: Exactly. It looks simple because it hides its decisions.

Mum *(from the kitchen)* He's right, you know. I've never understood why July and August are both long.

The Almanac - To understand how different calendars think about time, it helps to begin not with theory but with an object you can hold. In many Indian homes, that object is the traditional almanac, called a *panchang*, literally meaning "five parts." It is not designed for quick reference. It is designed to describe the sky. Pick one up and the first impression is overwhelming. The page for a single day lists the *tithi* - the lunar day; the *vara* - the day of the week; the *nakshatra* - the Moon's position among twenty-seven lunar mansions; the *yoga* - one of twenty-seven combinations defined by the relative positions of the Sun and Moon; and the *karana* - half of a lunar day. Alongside these appear planetary positions, sunrise and sunset times and any festivals or observances attached to the date. Six or seven distinct, overlapping cycles are presented together, all active at once, all treated as relevant.

The Illusion of Simplicity - The Gregorian calendar hides all of this. It shows you a single number - the date. Clean, unambiguous and deliberately stripped of complexity. But that simplicity is a design choice, not a feature of time itself. The Gregorian calendar was built for administration, for coordinating commerce, government and communication across large societies. It solved the problem of getting millions of people to agree on a single date, and it solved it well. In doing so, it set aside almost every other relationship between the calendar and the sky.

The complexity the Gregorian calendar hides has not gone away. Every lunisolar festival still follows its own cycle. Easter follows a rule combining the spring equinox, the first full moon after it and the first Sunday after that - three constraints that most people who celebrate it have never needed to work out. Chinese New Year is set by the second new moon after the winter solstice. Diwali falls on the new moon of the lunar month Kartik. The Gregorian calendar simply does not show any of this. It marks the result of these calculations as a fixed date on its grid and presents it as if the date were self-evident. This is why the Gregorian calendar appears simple. It was designed to be.

Two Different Purposes - A calendar that displayed all of these overlapping cycles at once would be difficult to use for most modern purposes. The *panchang* is not wrong. The Gregorian calendar is not wrong. They are answers to different questions, each optimised for a different way of relating to time. The *panchang* does not merely describe time. It also prescribes it. It indicates which hours are considered favourable for travel, which days suit new beginnings and which lunar phases carry particular significance. For hundreds of millions of people, this reflects a different and older relationship with the sky, one in which celestial movements do not simply mark time but help shape decisions within it.

The Three Cycles - Beneath these different systems lies a deeper constraint. Three distinct astronomical cycles govern time and none divides evenly into the others. This mismatch is the source of every calendar problem this book explores.

- The *day* is defined by the Earth's rotation. One full spin relative to the Sun takes twenty-four hours. This is the most fundamental unit, the one no civilisation has attempted to replace. The cycle of light and darkness is immediate and unavoidable.
- The *month* is defined by the Moon's cycle, roughly 29 and a half days from new moon to new moon. Twelve of these fall eleven days short of a solar year. Thirteen overshoot it. Neither aligns. The Moon does not move according to human preference.
- The *year* is defined by the Earth's orbit around the Sun. One full circuit takes about 365.2422 days. This, too, refuses to resolve neatly. It does not contain a whole number of days, nor a whole number of lunar months. A solar year spans approximately 12.3683 lunar months, a number that resists any attempt to be made whole.

An Unsolvable Problem - If the year contained exactly twelve lunar months, every culture might have arrived at the same calendar. If it were exactly 360 days, months could be fixed at thirty days each, and the system would be perfectly regular. The sky, however, does not produce convenient numbers. Every calendar is a response to this constraint. Some systems follow the Moon and set aside the Sun. Others follow the Sun and ignore the Moon. Still others attempt to balance both, inserting extra days or even entire months when the mismatch grows too large. The solutions differ widely, but the underlying problem is the same. A number like 12.3683 cannot be made whole, and every tradition must decide how to live with that fact.

This is not a problem that can be solved, only managed. No system based on whole days and months can perfectly align with both the solar year and the lunar cycle. For thousands of years, astronomers and calendar makers have developed ways to handle this limitation. Leap years, intercalary months, purely solar calendars and purely lunar ones are all strategies for working within the same constraint. Every calendar is, in the end, a negotiation with the sky.

Did You Know? The seven-day week has no astronomical basis. It was not derived from any cycle in the sky. It entered the Roman calendar from Babylonian tradition, spread through Jewish and Christian practice and has never been replaced. The other experiments of French ten-day week of 1793 and the Soviet five-day week of 1929 both failed within years. The seven day week is purely cultural and apparently immovable.

Story: The Emu in the Sky

Aboriginal Australia · tens of thousands of years

The fire has burned down to coals. The children have been asleep for hours.

An elder sits outside, looking up. The Milky Way stretches from one horizon to the other, dense with stars. But she is not watching the stars. She is watching the darkness between them.

There, running across the sky, the long neck, the rounded body, the trailing legs. The Emu. Not made of stars, but of the dark lanes of dust that cut through the galaxy, blotting the light behind them. You have to know how to look. She has known since she was younger than the children sleeping now.

She watches the Emu's posture. Tonight the head is low on the horizon just after sunset, the body stretching upward, the bird in full running stride. She nods slowly.

Tomorrow she will tell the others, the eggs are ready.

This is the calendar. Not a grid of squares on a wall, not a list of numbered days. The Emu's posture changes across the year as the Earth moves along its orbit, and each position marks something. When the eggs are in the ground, when the birds are nesting, when the rains are coming, when it is time for ceremony. The sky is not decoration. It is a working document, updated every night.

The knowledge lives in her the way all knowledge lives here. In the body, in the story, in the song that contains the song that was learned from the grandmother of her grandmother's grandmother. No clay tablet. No papyrus. No calendar hanging by a nail.

She looks at the Emu a moment longer.

Then she goes inside.

The Emu is still there. On any clear autumn night in the southern hemisphere, if you know where to look, you can find it running across the sky. The calendar it carries has been read on this continent for longer than any other calendar system on Earth has existed.

ARC 2
READING THE SKY

04 The Sky Was the First Clock

Another evening, a few days later. Dad has brought out a star chart - a large rotating disc that shows different star positions at different times of year. He holds it up toward the sky. Navya is trying to find the Pleiades and keeps pointing in the wrong direction.

Arjun: Before calendars were written down, how did people know what time of year it was?

Dad: They looked up. The sky told them everything. Which stars rose at dawn, how high the sun climbed at noon, where it set on the horizon.

Navya: *(pointing triumphantly at something)* There. The seven sisters.

Dad and Arjun both look. Neither is certain if she is right.

Arjun: How did they know which stars to watch? There are thousands.

Dad: Thousands of years of watching. The same patterns, the same appearances, the same disappearances every year. When you watch long enough the sky stops being decoration. It becomes information.

Navya: *(still pointing)* They're running.

Dad looks at her and smiles.

Dad: That's exactly what people saw in them. In many cultures, these stars were stories, people, animals. All moving across the sky.

Long before anyone wrote a calendar, every human community on Earth was doing the same thing. They were watching the sky and recording what they saw in memory, story and stone. The sky was not decoration. It was the primary source of information about time. What season it was, when floods were coming, when animals would migrate and when crops should be planted.

The Sun provided the most obvious signal. Its daily arc changes through the year. On the summer solstice it rises furthest north, climbs highest at noon and sets furthest north. On the winter solstice it barely climbs at all. A person who watched the sunrise position against the horizon every morning for several years would understand that this position moved in a perfectly predictable annual cycle. No instruments needed, just a fixed reference point and patience.

Stars provided a second layer of timekeeping. Every star rises approximately four minutes earlier each night because of Earth's orbital motion. As a result, different stars are visible at different times of year and the rising of specific stars at dawn or dusk became seasonal markers. The Pleiades are one of the most universally used stellar calendars in human history. Their reappearance before dawn in late spring is used as a seasonal signal in Aboriginal Australia, ancient Greece, ancient Egypt, among the Māori of New Zealand and across dozens of cultures in Asia and the Americas. The same cluster of stars, the same calendar signal, independently recognised on every inhabited continent.

Aboriginal Australians developed one of the most sophisticated sky calendars on Earth entirely without writing. The knowledge was encoded in stories containing precise astronomical information. When specific stars rose, how long they were visible and what ecological events their appearance signalled. The Boorong people of Victoria tracked not just stars but the dark spaces between them. The dust lanes of the Milky Way that form the Emu, a calendar figure invisible to every European astronomer who designed the world's major written calendar systems.

Around the World: The Pleiades star cluster is used as a seasonal calendar marker by cultures on every inhabited continent. Aboriginal Australians, ancient Greeks, ancient Egyptians, the Māori of New Zealand, the Aztecs of Mexico and dozens of communities across Asia and the Pacific. No other star cluster appears in as many independent calendar traditions.

05 The Moon That Made Months

The moon is wider now, four or five days older. Mum has come outside with tea and the family almanac, Panchang. She hands the almanac to Arjun without explaining why. He opens it to today's date and finds five columns of information he cannot yet read.

Arjun: *(looking at the Panchang)* What is all this?

Mum: *(sitting down)* Everything happening in the sky today.

Arjun: Why are months roughly thirty days?

Dad: Because that's how long the Moon takes from one new moon to the next. Ancient people noticed the cycle was perfectly regular and used it to count time.

Navya: *(reaching for the Panchang)* Can I see?

Arjun: *(not giving it up)* What's a nakshatra?

Dad: The Moon travels through the stars as well as going through its phases. It takes about 27 days to complete one full journey through the background of stars. Ancient Indian astronomers divided that journey into 27 stages and named each one. Each stage is a nakshatra. The almanac tells you which nakshatra the Moon is visiting today.

Arjun: *(finding the column)* Rohini. What does that mean?

Mum: *(with a slight smile)* Rohini is the most beautiful of the Moon's mansions. The Moon always lingers there.

Dad glances at Mum. She is not looking at him.

The Moon as Humanity's First Clock - Of all the astronomical patterns available to early humans, the Moon's cycle was the most accessible and the most immediately intelligible. Roughly every 30 days, the Moon completes its sequence of phases. From new moon through waxing crescent, first quarter, waxing gibbous, full moon, waning gibbous, last quarter, waning crescent and back again. No instruments are required, and no mathematics beyond counting. The pattern repeats with striking reliability, visible to anyone who looks up.

It is therefore not surprising that many of the earliest calendars were lunar. The Moon provided the most visible and countable rhythm in the night sky. The very idea of a "month" emerges from this relationship. Across languages, the connection persists. *Mensis* in Latin, *mah* in Persian, *masa* in Sanskrit, *yue* in Chinese. The Moon did not merely inspire the month. It defined it.

The Moon's cycle was not only visible but deeply entangled with the living world. Tidal patterns that governed coastal fishing followed its phases. Certain animals bred in alignment with lunar cycles. Some plants responded to changes in moonlight. A lunar calendar was therefore more than a way of marking time. It was a way of staying aligned with ecological rhythms that mattered for survival.

In the Vedic tradition, this observational knowledge was encoded in narrative. The Moon's twenty seven nightly positions were personified as his twenty seven wives, the daughters of Daksha, each associated with a *nakshatra* or star mansion. The Moon was said to visit each wife in turn, spending roughly one night with each before moving on. Beneath the story lies a precise observation. The Moon travels about 13 degrees each day, completing a circuit through the 27 nakshatras before repeating the cycle. The myth preserved the astronomy. A community that remembered the story was, in effect, tracking the Moon's motion across the sky. The system is documented in early Vedic texts, including the Atharvaveda.

The limitation of a purely lunar calendar, its steady drift against the solar year by about eleven days annually, is not immediately obvious. In societies without long written records, this discrepancy would have accumulated gradually across generations before becoming noticeable.

Reading Time in the Sky - Yet despite this limitation, the Moon offers something no other cycle does, it can be read directly. Every night the Moon appears different, but those differences are not random. They follow a precise and learnable pattern. With familiarity, the lunar date can be estimated from the sky itself, without consulting any written calendar.

This is possible because the Moon shines by reflecting sunlight. At any moment, half of it is illuminated and half remains in shadow. What changes is our viewing angle from Earth.

When the Moon lies between the Earth and the Sun, the illuminated half faces away from us. And we see the new moon, effectively invisible. As it moves along its orbit the lit portion grows, reaching half at first quarter and full opposite the Sun. It then shrinks back through the same sequence. Each phase rises and sets at a predictable time, determined by the Moon's position relative to the Sun.

Phase	Rises	Sets	What you see
New Moon	Sunrise	Sunset	Invisible. Lit side faces away from Earth. Moon rises and sets with Sun.
First Quarter	Noon	Midnight	Right half lit. High in the evening sky. Moon can be seen right over the head as the sun in setting.
Full Moon	Sunset	Sunrise	Fully lit. Opposite the Sun all night. Moon rises as the Sun is setting. Both full moon and sun can be seen on the other sides of horizon.
Third Quarter	Midnight	Noon	Left half lit. Visible before dawn

The sky itself carries the date for those who know how to read it. For most of human history, this was not specialised knowledge but practical skill. Fishermen watched the Moon to anticipate tides. Farmers observed it when planning planting cycles. Religious authorities relied on it to determine the start of a new month. In many traditions, the month does not begin at the astronomical new moon, which cannot be seen, but at the first visible crescent that follows. This requires direct observation. The calendar is confirmed by looking, not by calculation.

A Calendar That Requires Observation - In some traditions, this practice continues. The beginning of Ramadan, for example, is often marked by the sighting of the crescent Moon rather than by prediction alone. Witnesses must observe it directly. The month begins not when a calculation indicates its arrival, but when it is seen. It is a calendar that depends on human attention, one that still requires people to look up at the sky.

Story Board: The Moon's twenty-seven wives.
In Vedic tradition, the Moon's journey across the sky was imagined as a visit from one home to another. Daksha had twenty-seven daughters, each linked to a nakshatra, the star segments along the Moon's path. All were married to the Moon god Chandra, who was meant to spend equal time with each of them. But Chandra loved Rohini more than the others and lingered with her, returning often while the others waited. Hurt and angry, the sisters went to their father. Daksha cursed Chandra to fade and lose his light. As the Moon began to shrink, the gods intervened, and Shiva softened the curse. Chandra would not disappear, but wax and wane forever, visiting each wife in turn. The nakshatras in every almanac still trace that journey.

06 The Sun, the Year and the Wrong Season

A bright Saturday afternoon. Dad has pushed a stick into the soil in the sunny part of the garden. Its shadow falls clearly on the grass. Navya crouches beside it examining the shadow with interest.

Arjun: How did ancient people measure how long a year was? Before any instruments?

Dad: *(pointing at the stick)* This is an instrument. Put a vertical stick in the ground and measure its shadow at noon every day. The shadow changes length through the year. At the summer solstice it's shortest because the Sun is highest. At the winter solstice it's longest. Count the days between two summer solstices: 365.

Arjun: And they really did that?

Dad: They built magnificent versions of it. The Chinese built stone gnomons twelve metres high. The Egyptians aligned entire temples to catch the solstice sun. The principle is always the same: a shadow that tells you where you are in the year.

Navya: *(looking at the stick's shadow)* It's pointing south. *

Dad: At noon it always points true south in the southern hemisphere. In the northern hemisphere it points true north. Before the compass, that's how you oriented a building.

Arjun looks at the stick differently.

* The shadow points north in the Northern Hemisphere and south in the Southern Hemisphere.

Finding the Year - The length of the year, the time it takes for Earth to complete one orbit around the Sun, is not immediately obvious from direct observation. Days and lunar months are short enough to track directly. A full cycle unfolds within weeks. But a year spans 365 days and recognising that length requires either careful record keeping over multiple cycles or a fixed observation tied to a specific moment in the solar year.

One of the most reliable markers is the solstice. At the summer solstice, the Sun reaches its highest point in the sky at noon and the day is longest. At the winter solstice, it reaches its lowest point and the day is shortest. Even without instruments, these extremes are noticeable. The direction of sunrise and sunset shifts along the horizon. Shadows stretch or shrink dramatically. The change is gradual, but the turning points are unmistakable.

The equinox offers a more precise reference. On these days, the Sun rises very close to due east and sets very close to due west, and day and night are approximately equal in length. A simple vertical stick in the ground, a gnomon, is enough to track this. The length of its shadow at noon changes through the year, shortest at the summer solstice and longest at the winter solstice. By comparing these shadow lengths over time, early observers could identify key points in the solar cycle with surprising accuracy.

Different civilisations found different anchors. The Egyptians, for instance, looked not to shadows but to the stars. The heliacal rising of Sirius, its first appearance in the dawn sky after a period of invisibility, coincided with the annual flooding of the Nile. This provided a dependable marker of the year, tied to both the heavens and the rhythms of life on Earth. Different methods, but the same underlying discovery, the length of the year.

Why Seasons Exist - The reason the year matters lies in the seasons. Earth's axis is tilted about 23.5 degrees relative to its orbit around the Sun. This tilt, not the Earth's distance from the Sun, creates the cycle of seasons. When the northern hemisphere tilts toward the Sun in June, it receives more direct sunlight for longer hours. This is northern summer. When it tilts away in December, sunlight arrives at a lower angle for fewer hours. This is northern winter. The southern hemisphere experiences the exact opposite at the same time.

This reversal can feel counterintuitive because Earth is actually closest to the Sun in early January, during northern winter, and farthest in early July, during northern summer. The difference in distance is small, about three percent, and is overwhelmed by the effect of tilt and daylight duration. Seasons are governed by geometry, not proximity. What matters is not how far you are from the Sun, but how directly its light reaches you and for how long.

When Calendars Travel - Most of the world's major festival traditions took shape in the northern hemisphere, where the majority of the global population has historically lived. Their timing reflects the ecological and seasonal rhythms of that region. Diwali is linked to the north Indian autumn, after the monsoon and before winter. Chinese New Year marks the transition toward northern spring. Easter falls at the end of northern winter. Christmas arrives in the depth of northern midwinter.

When these calendars move across hemispheres, their seasonal meaning shifts. In countries such as South Africa, Argentina and New Zealand, the same festivals occur under entirely different conditions. Diwali falls in spring rather than autumn. Chinese New Year arrives in late summer rather than late winter. Easter appears as summer fades, and Christmas takes place in the height of southern summer.

The astronomical timing remains unchanged. The Moon's phases look the same from Buenos Aires as from Mumbai. But the seasonal context is reversed. Some communities preserve the original associations, treating the inversion as part of life in a different landscape. Others adapt gradually, blending local seasonal elements into inherited traditions. Spring flowers may accompany Diwali. Christmas may take on the character of a summer festival.

Neither approach is incorrect. Both reflect what happens when a calendar travels beyond the environment in which it was first shaped.

Strange But True: Stonehenge was built over 5,000 years ago with no written language, no metal tools and no wheeled vehicles. Its massive stones are aligned so precisely that at the summer solstice sunrise, the Sun rises directly over the Heel Stone and lights the centre. The people who built it did not leave a single written word, but they left this. Across the world, other civilisations did the same. At Chichén Itzá in Mexico, the setting Sun at the equinox casts a moving shadow that looks like a serpent descending the pyramid. At Angkor Wat in Cambodia, the towers align with the rising Sun at key points of the year. In ancient Egypt, temples were built so that sunlight would reach inner sanctuaries only on specific days. Again and again, without writing, people marked the solstices and equinoxes with astonishing precision. The sky was their calendar, and they read it well.

Around the World: In the southern hemisphere Christmas falls in midsummer. South Africans, Argentinians, Australians and New Zealanders celebrate it in December heat with outdoor gatherings, summer fruits and long bright evenings. The date is fixed by the solar calendar and cannot move. Only the meaning of December has changed.

07 When the Pole Stars Change: Precession

Late evening, darker. Dad has the star chart out again, pointing at Polaris. Arjun has found it. Navya is lying on her back looking at the whole sky at once.

Arjun: The Pole Star is always north. Right?

Dad: Right now it is. Polaris sits almost exactly above Earth's North Pole at this moment.

Arjun: At this moment?

Dad: Earth's axis wobbles. Very slowly. One full wobble takes 26,000 years. But it means the direction the pole points changes over millennia. Right now it points at Polaris. Five thousand years ago, when the Egyptian pyramids were being built, the pole star was Thuban. In the constellation Draco.

Navya: *(from the grass)* Thuban.

Dad: The pyramid builders had a completely different North Star. They built certain shafts in the pyramids to align with Thuban. Those same shafts no longer point at any pole star.

Arjun: And in the future?

Dad: In 14,000 years the pole star will be Vega. Much brighter than Polaris. Future navigators will have a far more conspicuous guide.

Arjun: So every ancient astronomical observation is describing a slightly different sky from the one we see tonight.

Dad: Every single one.

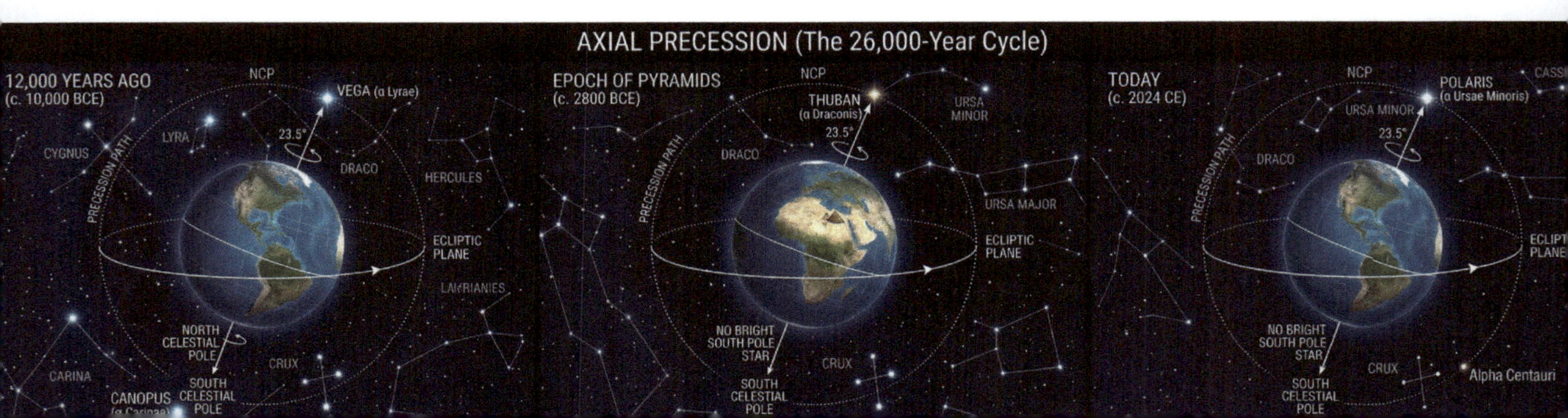

Earth spins on an axis tilted 23.5 degrees from vertical. Like a spinning top knocked slightly off centre, Earth's axis does not stay fixed in space. It traces a slow circle through the sky. This wobble is called precession and it takes approximately 26,000 years to complete one full circle. It is one of the most consequential movements in astronomy and one of the least discussed in everyday life.

The practical consequence is that the direction the North Pole points changes over millennia. At present it points almost exactly at Polaris, the familiar North Star. But 5,000 years ago the pole was closest to Thuban, a star in the constellation Draco. The builders of the pyramids at Giza are thought to have used Thuban as their celestial north, and certain shafts in the pyramids are believed to align closely with its position around 2500 BCE. Those same shafts point at empty sky today.

In approximately 14,000 years the bright star Vega will become the northern pole star, which is far more conspicuous than Polaris and one of the brightest stars in the northern sky. In 26,000 years the cycle will complete and Polaris will return to approximately its current position.

Precession also shifts the position of the spring equinox against the background of stars at about one degree every 72 years. Over 2,000 years this has moved the equinox point by nearly one full zodiac sign. This is why astrological star signs no longer correspond to the constellation the Sun is actually in at your birth. The signs were fixed to the sky two millennia ago and the sky has since moved on.

For calendar-makers the consequences are concrete. The tropical year measures from one vernal equinox to the next - 365 days, 5 hours and 48 minutes. The sidereal year measures the Sun's return to the same position against the fixed stars - 365 days, 6 hours and 9 minutes. The difference is 20 minutes and 24 seconds per year. Every 72 years this accumulates to one day.

Indian solar calendars are sidereal - they track the Sun against the actual stars. The Gregorian calendar is tropical - it tracks the Sun against the equinox point. Because precession moves the equinox point slowly backward through the stars, the two systems drift apart by one day every 72 years. The Gregorian vernal equinox falls on March 21st. Mesha Sankranti, the Indian solar new year, falls on April 14th. That 24-day gap is approximately 1,700 years of accumulated drift since the two systems were last aligned around 285 CE.

Sky Note: The southern hemisphere has no pole star. The south celestial pole sits in empty sky. Southern navigators used the Southern Cross to estimate south and is the reason it appears on the flags of Australia, New Zealand, Brazil, Papua New Guinea and Samoa.

Story: Arundhati and Vashishtha: The Stars That Remember
Indian astronomy · Ramayana · approximately 11,000 BCE (debated)

The wedding is over. The fire has been circled, the vows made, the rice thrown. Now the groom leads his new wife away from the lamplight, away from the guests, to a quiet patch of dark ground where the sky is visible.

He raises his arm and points."Vashishtha," he says. A star, steady and bright. "And there, just beside him, do you see? Arundhati. His wife."

She looks. Two stars, close together. The fainter one beside the brighter.

"She is always near him," the groom says. "Always faithful. Always in view." He is reciting something he has been taught, words passed down through more generations than either of them can count. "May you be like Arundhati."

She does not say what she is thinking, that in the Ramayana, the great text, Arundhati does not follow Vashishtha. She walks ahead of him. The text notes it as something remarkable, worth recording. The smaller star, the wife, leading the larger star, the sage.

That is not what the sky shows tonight. Tonight Arundhati follows.

She will ask someone about this later. A priest, or perhaps a scholar of the ancient texts. The answer, if she ever finds it, will be extraordinary. That the Earth's axis wobbles over a cycle of 26,000 years, and that there was a time, thousands of years before any written record, when the sky looked different enough that Alcor did appear to precede Mizar across the night.

The observation was made. It was put into a story about a marriage. And it survived, inside that story, for thirteen thousand years.

The groom is still looking at the stars. She takes his arm.

The wedding ceremony still exists. Grooms still point to those two stars every day somewhere in the world, still recite the same words, still ask their brides to be like Arundhati. Nobody performing the ceremony knows they may be repeating an astronomical observation from thirteen thousand years ago. The story remembered it for them.

Story: The Priests of Sirius
Egypt · Old Kingdom · approximately 2500 BCE

An hour before dawn. The temple roof. The priest faces east and waits.

For seventy days Sirius has been gone. Not dead. Just lost in the Sun's glare, too close in the sky to be visible. Every morning for seventy mornings he has climbed here and found nothing. But the count is right. Tonight is the night, or close to it.

He steadies his breathing. Watches the horizon.

Then, so brief he almost doubts it. A spark of blue-white light at the very edge of the world. Gone in an instant as the sky begins to pale. But he saw it.

He raises his hand. His assistant, waiting behind him, makes the mark.

Sirius has risen. Sopdet, the goddess of the flood. The star the Egyptians have been watching every year since before the pyramids were built, since before anyone now living can remember there being a before. When Sirius rises here, the Nile rises within weeks. Not because Sirius causes the flood. The priest knows better than that. But because both things happen at the same time, every year, reliably, the way all true things in the sky are reliable.

He turns and walks down from the roof. The calculation is done. The new year has begun.

In 46 BCE, two thousand years from now, a Roman general named Julius Caesar will borrow this 365-day year from Egypt and impose it on the Western world. In 1582 CE a Jesuit mathematician will refine it. The calendar on a wall in a house in Melbourne will be the direct descendant of the number this priest has just confirmed on a rooftop beside the Nile.

He does not know any of that. He goes inside for breakfast.

The calendar the priest confirmed on that rooftop is the one most of the world uses today. It passed through Julius Caesar, through Pope Gregory XIII and through Christopher Clavius before arriving on your wall. You probably checked it this morning without thinking about the priest who made it possible, or the star that told him when to begin.

ARC 3
BUILDING CALENDARS

08 Egypt and the 365-Day Year

Weekend morning. Dad has a library book on the table, open to a photograph of an Egyptian star clock. A circular disc with thirty-six figures around the rim. Arjun is reading it over breakfast. Navya wanders out holding toast.

Arjun: Who made the first proper solar calendar?

Dad: Egypt. About five thousand years ago. Priests discovered that the star Sirius rose just before dawn at almost exactly the same time each year, just before the Nile flooded. They counted the days between two risings, 365 days.

Arjun: But 365 isn't quite right. The year is 365.25 days. So the Egyptian calendar was losing a quarter day every year.

Dad: Yes. They were aware of the drift over time and tracked it carefully, even if they did not express it in the exact fractional terms we use today.

Arjun: Why didn't they fix it sooner?

Dad: Changing the official calendar would have disrupted every religious festival tied to it. The priests knew the drift was there. They tracked it carefully. But the cost of changing was high. So they lived with the drift. Their year had 12 months of 30 days each, plus five extra days added at the end.

Navya: *(still holding her toast)* What were the five extra days at the end?

Dad: Considered unlucky. Outside normal time. No government business, no weddings, no contracts. The Egyptians called them the days upon the year. Days that didn't quite belong.

Navya: *(thoughtfully)* I want some days that don't quite belong.

Think about what a calendar actually needs to do. It must divide the year into manageable units, give those units consistent lengths and anchor the whole structure to something reliable in the sky. Around 3000 BCE the Egyptians did all three in one stroke, and the structure they invented has never fundamentally changed.

They divided the year into twelve months of exactly thirty days each. Three hundred and sixty days. Then they added five extra days at the end, bringing the total to 365. This was the first time any civilisation had imposed a clean, repeating twelve-month structure on the solar year. Every calendar that followed, including the one on your wall today, is built on this same foundation. Twelve months, approximately thirty days each, totalling 365.

The Egyptians anchored their year to the heliacal rising of Sirius, the brightest star in the sky, which reliably appeared before dawn just before the annual Nile flood. The New Year began on that day. The calendar then ran its 365 days and the cycle began again. It was elegant, stable and slightly wrong. The true tropical year is 365.2422 days, meaning the Egyptian calendar lost approximately one day every four years. Over centuries this accumulated into a visible drift. The Egyptians observed and tracked this drift carefully, developing a concept later known as the Sothic cycle, a period of 1,460 years for the calendar to return to alignment with Sirius. They knew exactly how wrong they were and chose to live with it rather than disrupt the religious ceremonies anchored to every date in the year.

The five extra days at the end were treated as a threshold. Time outside normal time, belonging to neither the old year nor the new. No official business. No marriages. No contracts. Days suspended between one year and the next.

The Egyptian calendar endured in continuous use for over three thousand years. No civilisation before them had managed to impose such a clean, stable structure on the solar year. No civilisation since has fundamentally altered that structure. The twelve-month, 365-day framework that Egyptian priests built around the rising of Sirius is the skeleton of every solar calendar that followed. The calendar that Egypt invented beside the Nile is the calendar that most of the world woke up to this morning.

Did You Know? The Egyptians believed the five extra days were so outside normal time that not even gods could be born during the regular year. They reserved these threshold days for their most powerful deities. Osiris the god of the dead, Isis the goddess of magic, Horus the sky god, Set the god of chaos and Nephthys the goddess of the night. The regular calendar, it seems, could not contain them.

09 The Vernal Equinox: The Universal Anchor

Late afternoon. The light casts equal shadows on both sides of the fence post in the garden. Navya is lying in the grass watching the shadows. She has been watching them for a while.

Arjun: Is there any single astronomical event that matters to lots of different calendars at the same time?

Dad: Yes. The vernal equinox. The spring equinox in the northern hemisphere, around March 20th or 21st. Day and night are exactly equal. The Sun rises exactly due east and sets exactly due west.

Arjun: Why would that matter so much to so many different cultures?

Dad: Because it is the moment winter ends. Day begins to overtake night. The soil warms. Seeds germinate. Animals give birth. Every farming civilisation on Earth has watched this moment and felt something shift. Before astronomy, before writing, the return of spring was the most important event of the year.

Navya: *(still watching the shadows)* The shadows are the same on both sides right now.

Dad and Arjun both look at the fence post.

Dad: *(quietly)* They are. It's almost equinox.

Navya nods as if this confirms something she already knew.

Arjun: So calendars anchored to the equinox are really anchored to spring itself.

Dad: Exactly. And civilisations that built their new year around it were saying, this is where time begins. Not because of politics or convenience. Because of what the earth actually does.

The vernal equinox is the moment in spring when the Sun crosses the celestial equator moving northward. Day and night are exactly equal, the Sun rises precisely due east and sets precisely due west. But its significance to the world's calendars is not merely astronomical. It is ecological.

At the vernal equinox, winter ends. Days begin to lengthen past nights. Soil that has been cold for months starts to warm. Seeds that have been dormant begin to germinate. Birds return. Animals give birth. The entire natural world responds to this precise astronomical moment. For any farming civilisation watching the land, it was the most important day of the year. The day life restarted.

This is why cultures that had no contact with each other, separated by oceans and mountain ranges and thousands of years, all arrived at the same astronomical anchor independently. Nowruz, the Persian New Year, falls exactly on the vernal equinox, calculated to the hour. It has been celebrated for over three thousand years, through empires, conquests and conversions. The haft-seen table is set at the exact instant of the equinox. Not the morning of, not the evening before, but the moment itself.

Across the ancient world, Stonehenge, the temples of Egypt, the pyramids of Mesoamerica, stone circles across Europe and Asia, the equinox was carved into stone. People who could not write built structures aligned to the day the light returned. That impulse is older than any calendar.

The equinox is not simply an astronomical convenience. It is a threshold. It is the same moment in Earth's orbit felt by Persian farmers setting the haft-seen table, by Aboriginal Australians reading the emu's posture, by the priests who aligned Stonehenge to catch the light. Different names, different ceremonies, different calendars. The same turning point in the sky.

Around the World: Nowruz, meaning "new day" in Persian, is celebrated by more than 300 million people across Iran, Afghanistan, Tajikistan, Azerbaijan, Kurdistan and the wider diaspora. It marks the exact moment of the spring equinox, not a day or a morning, but the precise second the Earth tilts toward spring. Families prepare the *haft-seen* table, which means "seven S's", with seven symbolic items beginning with the Persian letter S, each representing renewal, health, prosperity and life. They gather and wait for that moment, confirmed by astronomical calculation. Nowruz is more than 2,000 years old, older than Islam itself, and in 2009 it was recognised by UNESCO as an Intangible Cultural Heritage of Humanity.

10 Rome's Calendar Catastrophe

A warm evening, windows open. Dad has drawn a timeline on the back of an envelope. 304, then 355, then 445, then 365. The number 445 is circled. The envelope is slightly crumpled because he drew it in his pocket earlier and forgot about it. Navya picks it up and examines the circled number.

Arjun: How bad was the Roman calendar before Julius Caesar fixed it?

Dad: *(taking the envelope back)* The original Roman calendar had 304 days and ten months. Which means they simply didn't count winter. January and February didn't exist.

Arjun: They skipped winter.

Dad: Officially. The year ended in December and resumed in March.

Navya: *(with quiet longing)* I want uncounted time.

Arjun: So why do some months have 29 days and some 31? Why not just make them all 30?

Dad: Superstition. Romans believed odd numbers were lucky and even numbers unlucky. When King Numa redesigned the calendar around 713 BCE he deliberately gave most months an odd number of days. February got 28, an even number, because February was already the month of the dead. An unlucky month could have an unlucky number.

Arjun: *(slowly)* So our months are unequal because of a superstition from 700 BCE.

Dad: That, and one emperor's pride later on. By the time Caesar arrived the calendar had been manipulated for so long that it was running three months ahead of the actual seasons. He had to insert 445 days into a single year just to reset everything.

Navya: *(looking at the circled 445)* That's almost a year and a half.

Dad: Exactly. The Year of Confusion. And even after the reset, the priests who inherited Caesar's calendar immediately misread his leap year rule. Rome could not leave a calendar alone.

After dinner. Dad has taken the wall calendar down and laid it flat on the table. He is looking at something. Arjun leans over to see what.

Arjun: Why do July and August both have 31 days? Every other pair of adjacent months alternates between long and short.

Dad: That is the result of two emperors, one bureaucratic blunder and one patient correction.

Arjun: What do you mean?

Dad: After Caesar was assassinated, the priests misread his leap year rule. They inserted a leap year every three years instead of every four. Augustus discovered the error and corrected it quietly by suspending leap years for several decades until the drift drained away.

Navya *(from the doorway):* February was robbed.

Dad: *(without turning around)* That's what people often say, though the history is not quite so simple. Once the calendar was running correctly, Augustus renamed the month Sextilis after himself and insisted it have 31 days like Julius's July. He took the extra day from February.

Arjun: And why does the year start on January 1st? That seems arbitrary too.

Dad: It is. January is named after Janus, the Roman god of doorways, who has two faces. One looking back at the old year and one looking forward to the new. Numa Pompilius, the second king of Rome, placed January first for that reason. But the specific date of January 1st was Julius Caesar's choice. That was when Roman consuls took office.

Arjun: And the year number itself? Who decided we are in this particular year?

Dad: A sixth-century monk called Dionysius Exiguus, roughly, Dennis the Small. He created the BC and AD system to standardise Easter calculations. He set the birth of Jesus as Year 1. Most historians think he got it wrong by at least four years. And he forgot to include a year zero, which creates a mathematical problem astronomers still work around today.

Arjun: He forgot zero?

Dad: He forgot zero.

Navya *(still in the doorway)* I feel sorry for the year that doesn't exist.

The Roman calendar was, for most of its history, a demonstration of what happens when administrative convenience overrides astronomical reality for long enough. Every ruler who touched it left a mark. Almost none of those marks were astronomical.

The original calendar had ten months and 304 days. Winter was simply not counted. The year ended in December and resumed in March as though the cold months between had no official existence. It was a calendar built for convenience, not accuracy, and it began drifting from its first year.

By the time King Numa Pompilius reformed it around 713 BCE, adding January and February to bring the year to 355 days, the Romans had a lunar calendar that fell behind the solar year by about ten days annually. They attempted to correct this by inserting an extra month, Mercedonius, in alternate years. The mechanism was sound. The problem was who controlled it.

Intercalation was the responsibility of the College of Pontiffs, Rome's most powerful religious body. They used it as a political instrument. A year could be extended to keep a favoured magistrate in office. A year could be shortened to force an opponent's term to end early. The calendar was not tracking the sky. It was tracking ambition.

By the 40s BCE Julius Caesar became dictator, the calendar was running approximately three months ahead of the actual seasons. October on paper, autumn in the fields. He turned to Egypt for help. Working with Sosigenes of Alexandria, one of the finest mathematical astronomers of the ancient world, Caesar designed the most rational calendar Rome had ever attempted. Sosigenes brought the Egyptian precision of the 365-day year and the calculation of the quarter-day remainder. Together they settled on 365 days per year with a leap day every four years.

But first Caesar had to reset the damage. He declared 46 BCE a year of 445 days, inserting three extra months to drag the calendar back into alignment with the actual seasons. The Year of Confusion. Then the Julian calendar began on January 1st, 45 BCE.

Two years later Caesar was assassinated. He did not live to see his own reform take hold.

Did You Know? The Year of Confusion, 46 BCE, is the longest year in Western calendar history at 445 days. Julius Caesar inserted three extra months to reset a calendar three months out of step with the seasons. The largest single calendar correction ever made in one move.

The Julian calendar died almost immediately after its creator. Caesar was assassinated in 44 BCE. The priests who inherited his calendar misread his leap year rule. Caesar had specified a leap year after every three year, but the priests counted inclusively and inserted one every third year. Between 45 BCE and 9 BCE they inserted twelve leap years when only nine were correct. Augustus discovered the error and corrected it with characteristic patience. Where Caesar had disrupted everything dramatically, Augustus suspended leap years for several decades and let the error drain away quietly. Nobody noticed it happening. The correction completed in 8 CE.

Once the calendar was running correctly, Augustus turned to the question of his legacy. He had watched Caesar name a month after himself. The Senate obligingly renamed the month Sextilis in Augustus's honour. But Augustus found it intolerable that his month should have only 30 days while Caesar's July had 31. He demanded an extra day. February, which had no emperor's name and no political protector, was the obvious source. The cascade of adjustments this required, shortening of September and November and lengthening of October and December. This is visible in the final column of the table that follows next.

January is named after Janus, the Roman god of doorways, two-faced, looking back at the old year and forward to the new. Numa Pompilius placed January first around 713 BCE with that mythological reasoning. Julius Caesar then fixed January 1st as New Year's Day because that was when Roman consuls took office. A political choice with no astronomical basis. The month names confirm the same pattern. September, October, November and December still carry Latin numbers for seven, eight, nine and ten while sitting in positions nine through twelve. The calendar was reorganised by politics and nobody updated the labels.

The year number carries one final arbitrary decision. A sixth-century monk named Dionysius Exiguus created the BC and AD system to standardise Easter calculations. He set the birth of Jesus as Year 1. Most historians believe he miscalculated by at least four years. He also omitted year zero entirely, going directly from 1 BCE to 1 CE. Every year number we use today carries his arithmetic, his error and his missing zero forward permanently.

Strange But True: Because Dionysius Exiguus did not include a year zero when numbering years, the BC/AD system moves directly from 1 BCE to 1 CE. As a result, each century and millennium begins with a year ending in "1" and ends with a year ending in "00". This means the first millennium ended on December 31, 1000 CE, and the second millennium began on January 1, 1001 CE, ending later on December 31, 2000 CE. One year later than the popular celebrations that marked their turning points.

The Roman Calendar: How It Changed

Each column below shows the calendar as it stood under a different ruler. The month lengths are not arbitrary. Each change reflects a specific political or superstitious decision.

Month	Romulus (750 BCE)	Numa (713 BCE)	Julius Caesar (46 BCE)	Augustus (8 BCE)
Januarius	-	29	31	31
Februarius	-	**28**	29	**28**
Martius	31	31	31	31
Aprilis	30	29	30	30
Maius	31	31	31	31
Iunius	30	29	30	30
Quintilis / July	31	31	**31**	31
Sextilis / August	30	29	30	**31**
September	30	29	31	30
October	31	31	30	31
November	30	29	31	30
December	30	29	30	31
Total	304	355	365	365

Romulus : 10 months of 31 and 30 days with no counting of Winter months

Numa : All odd months because of superstition. Only Feb as even to make the entire year odd. Aligns with 11 Lunar cycle of 354.5 days

Julius : Months of 31 and 30 days with February as 29 days. Quintilis changed to July after himself.

Augustus : Five Months changed to accommodate change in August

The Roman months reflect a mix of gods, festivals and numbers. Januarius from Janus, god of doorways and beginnings. Februarius from Februa, a purification festival. Martius from Mars, god of war; Aprilis of uncertain origin. Maius from Maia, goddess of growth and Iunius from Juno, queen of the gods. Quintilis (fifth) and Sextilis (sixth) were later renamed July after Julius Caesar in 44 BCE and August after Augustus in 8 BCE. The remaining months retain their numerical origins. September (seven), October (eight), November (nine) and December (ten).

Romulus, 750 BCE: 304 days Rome's first calendar had ten months beginning in March and ending in December. The roughly 61 days of winter between December and March simply did not exist in the official record. No dates, no names, no count. Modern historians debate whether this reflects genuine indifference to winter or a deliberate decision that winter had no civic or agricultural significance worth marking. Either way, Rome began its calendar life by refusing to acknowledge nearly two months of every year.

Numa Pompilius, 713 BCE: 355 days The second king of Rome added January and February to fill the winter gap, bringing the year to 355 days. But Numa did not simply add 61 days wherever they fit. He redesigned the existing months guided by a Roman superstition that ran deep. Odd numbers were auspicious and even numbers inauspicious. He changed most 30-day months to 29 days so they would carry an odd count. February was the exception, given 28 days, an even and unlucky number, because February was already the month of purification rituals and the dead. An inauspicious month could carry an inauspicious number. The unequal month lengths visible in the table today begin with this decision, made for reasons that had nothing to do with the sky.

Julius Caesar, 46 BCE: 365 days By Caesar's time the calendar was three months ahead of the actual seasons, corrupted by decades of political intercalation. Caesar declared 46 BCE a year of 445 days, inserting three extra months to reset the drift completely. He then established 365 days with a leap day every four years. The most accurate civil calendar Rome had ever managed. He was assassinated two years later before his reform had fully taken hold.

Augustus Caesar, 8 BCE: 365 days Augustus inherited a calendar still carrying a priestly error. His predecessors had misread Caesar's leap year rule. He quietly suspended leap years for several decades until the error drained away. Then, with the calendar running correctly, he renamed the month Sextilis after himself. He insisted it have 31 days, matching Julius Caesar's July. February, with the least political protection of any month, lost the extra day. The reshuffle cascaded through September, October, November and December. Five months moved to accommodate them.

11 The Man Who Fixed the Calendar

After the dishes. Dad has placed a library book open at the centre of the kitchen table. A portrait of Christopher Clavius, bearded, in Jesuit robes, holding a pair of mathematical dividers. Arjun studies the portrait.

Arjun: Who actually designed the Gregorian calendar?

Dad: The reform was based on the proposal of the Italian astronomer Luigi Lilio, with the Jesuit mathematician Christopher Clavius playing a central role in explaining, refining and defending it. He was considered one of the finest mathematicians in Catholic Europe in the sixteenth century.

Arjun: And what was the problem he was solving?

Dad: The Julian calendar had accumulated ten days of drift since 325 CE. The spring equinox was happening on March 11th instead of March 21st. Easter was drifting away from spring. The Pope appointed a commission. Clavius led the mathematical work.

Arjun: Ten years for a calendar?

Dad: The mathematics took months. The politics took a decade. Convincing the Catholic world to accept that ten days would simply be removed from October required considerably more than good arithmetic.

Navya *(looks at the portrait)* He looks like he knew he was right from the beginning.

Dad *(studies the portrait)* He probably did. He spent the last years of his life writing defences of the reform against critics who refused to accept it.

By the late sixteenth century, the Julian calendar had accumulated approximately ten days of drift from the true tropical year. The problem was structural. The Julian leap year rule added a full quarter day every four years, but the true tropical year is 365.2422 days, not 365.25. The difference is only eleven minutes annually, but eleven minutes compounded over 1,250 years becomes ten days. By 1582 the spring equinox, which the Council of Nicaea in 325 CE had placed on March 21st for the purpose of calculating Easter, was arriving around March 11th. The Council of Trent had already described the situation as a scandal in 1563. Pope Gregory XIII assembled a commission to fix it.

Christopher Clavius, a German Jesuit holding the chair of mathematics at the Collegio Romano in Rome, led the astronomical work. He was not the most famous mathematician of his age, that was Galileo. He was not the most celebrated astronomer, that was Tycho Brahe. He was something rarer, a man who combined mathematical precision with institutional patience. The reform required not just a correct solution but one that could be explained to every bishop, clerk and calendar printer across Catholic Europe.

Clavius faced two distinct problems. The first was the accumulated drift, ten days needed to be removed from the calendar immediately. The second was the future drift, the Julian rule needed to be tightened to prevent the same error recurring.

His solution to the first problem was blunt and effective. Ten days were simply removed from October 1582. The day after October 4th became October 15th. Contracts, birthdays and feast days that fell in those ten days had no date that year. Clavius acknowledged this would cause disruption and decided the disruption was less costly than continued drift.

His solution to the future problem was mathematically elegant. He kept the four-year leap year rule but removed three leap years from every 400-year period. Century years would not be leap years unless they were also divisible by 400. This three-layer rule produces an average year of 365.2425 days, within 26 seconds of the true tropical year. The Gregorian calendar will not require a comparable correction for approximately 3,000 years.

Worth Knowing: Galileo Galilei, who would later clash with the Catholic Church on major questions of cosmology, travelled to Rome in 1611 to demonstrate his telescope and its discoveries, seeking out Christopher Clavius, the leading architect of the Gregorian calendar reform. Clavius and his fellow Jesuit scholars examined the evidence, including the moons of Jupiter, and confirmed that what Galileo was seeing was real rather than an optical illusion. Clavius died in 1612 still defending the calendar reform he had designed.

12 The Missing Days: Ten Days Erased

Dad has written out a calendar page for October 1582 on a large piece of card. October 1 through 4, then a wide gap, then October 15 through 31. He leans it upright against the fence. Everyone looks at the gap.

Arjun: October 4th was followed by October 15th. Ten days simply didn't exist.

Dad: For everyone under papal jurisdiction. Yes.

Arjun: What happened to things scheduled in those ten days? Contracts. Rent payments. Feast days.

Dad: Courts were busy for months. Landlords argued tenants owed ten more days of rent because the days had been skipped. Tenants argued the days had never happened. Saints whose feast days fell in the gap were simply moved or went unobserved.

Navya: *(tracing the gap with her finger)* What happened to people who were born then?

Dad: In Catholic countries, that birthday ceased to exist. In Protestant countries that hadn't adopted the reform yet, it continued normally. Europe had two different dates for the same physical day.

Arjun: For how long?

Dad: 170 years in Britain's case. Every international letter and contract had to specify Old Style or New Style to be understood.

CALENDARIVM ANNI MDLXXXII – REFORMATIO GREGORIANA
Calendar of 1582 – Gregorian Reform

September

Mo	Tu	We	Th	Fr	Sa	Su
					1	2
3	4	5	6	7	8	9
10	11	12	13	14	15	16
17	18	19	20	21	22	23
24	25	26	27	28	29	30

October

Mo	Tu	We	Th	Fr	Sa	Su
1	2	3	4	15	16	17
18	19	20	21	22	23	24
25	26	27	28	29	30	31

November

Mo	Tu	We	Th	Fr	Sa	Su
1	2	3	4	5	6	7
8	9	10	11	12	13	14
15	16	17	18	19	20	21
22	23	24	25	26	27	28
29	30					

December

Mo	Tu	We	Th	Fr	Sa	Su
				1	2	3
4	5					
6	7	8	9	10	11	12
13	14	15	16	17	18	19
20	21	22	23	24	25	26
27	28	29	30	31		

Adoption of the Gregorian Calendar by Papal Bull of Pope Gregory XIII

On the evening of Thursday October 4th 1582 the Julian calendar ended in Catholic Europe. The following morning was not Friday October 5th. It was Friday October 15th. Ten days had been removed overnight. A correction for the ten days of drift that had accumulated since the Council of Nicaea in 325 CE. The correction was mathematically precise and historically extraordinary. No comparable event had occurred since the Roman calendar's Year of Confusion in 46 BCE.

The practical consequences were immediate and deeply disruptive. Contracts specifying payment between October 5th and 14th referred to dates that no longer existed. Estate leases, loan agreements and promissory notes that spanned the removed days became subjects of urgent legal dispute. Interest calculations were contested. Courts across Catholic Europe dealt with the fallout for months. The Spanish crown documented numerous commercial disputes in its legal archives from the weeks following the reform. Nobody had considered what happened to obligations falling in a gap between two calendar systems.

The most haunting individual case occurred at the precise hinge of the reform. Teresa of Ávila, the Spanish mystic and reformer of the Carmelite order, died on the night of October 4th 1582, the last night of the Julian calendar. She was buried the following day, October 15th, the first morning of the Gregorian one. Her death certificate records October 4th. Her feast day is October 15th. She is the only major historical figure whose death and burial fall on either side of the ten missing days, with nothing in between.

Protestant countries rejected the reform, viewing it as a Catholic imposition requiring no compliance. The split had immediate practical consequences. Every international letter and contract had to specify Old Style or New Style to avoid confusion. A date written in London and read in Madrid referred to different days. Merchants, diplomats and scientists working across the divide carried the discrepancy constantly. Britain did not remove its eleven days until September 1752. Russia waited until February 1918, by which point the Julian calendar was thirteen days behind. For nearly three centuries, Europe ran two simultaneous calendar systems.

A commonly repeated story says that at the 1908 London Olympics, Russian athletes arrived late because Russia was still using the Julian calendar, which by then ran thirteen days behind the Gregorian. Some historians, however, question whether this actually happened.

True Story: When Britain removed eleven days in September 1752, Parliament had to legislate about money specifically. The Calendar New Style Act required that rents, debts and legal deadlines falling in the removed days be deferred by eleven days. Otherwise, landlords could demand payment for time that had never legally existed. The British tax year was also shifted to avoid losing eleven days of revenue. It runs from April 6th to this day. A direct consequence of a calendar reform 270 years ago.

13 The 24-Hour Day: Where Hours Come From

After dinner. Arjun is looking at the clock on the kitchen wall. He turns around.

Arjun: Why are there 24 hours in a day? The day comes from Earth's rotation. The year from its orbit. The month from the Moon. But 24, where did that come from?

Dad *(thinks for a moment)*: Egypt.

Arjun: Egypt invented the hour?

Dad: Egypt invented the division of the day. They split daylight into ten hours, added a twilight hour at each end, then divided the night by watching which stars rose after dark. Twelve stars, twelve night hours. Ten plus two plus twelve, twenty-four.

Navya: *(from the floor, without looking up)* That is a lot of counting stars.

Dad: Every night. For centuries. The same twelve stars rising in sequence, one per hour. They called them the decans.

Arjun: And the hours were equal?

Dad: Not at first. Egyptian hours stretched in summer and shrank in winter. Equal hours came from Babylon. They divided the full day into 24 equal parts regardless of season. That is the hour we use today.

Navya: So our clocks run on Babylonian time.

Dad: In a way. Yes.

The day, Earth's single rotation, is the most fundamental unit of time. But the decision to divide it into 24 equal hours is not natural or inevitable. It is a specific invention with a specific origin, and like so much in the history of calendars, it arrived through two different civilisations solving the same problem in different ways.

The Egyptians began with the Sun. They divided daylight into ten observable hours and added a twilight hour at each end for a total of twelve. For the night they used a different system, the decans. These were thirty-six groups of stars whose heliacal risings they tracked across the year. On any given night, twelve of these star groups rose in sequence across the sky, one per hour, from dusk to dawn. Twelve night hours plus twelve day hours, twenty-four.

Egyptian hours were not equal in length. Because daylight is longer in summer and shorter in winter, each Egyptian hour stretched or shrank with the season. A summer daytime hour was considerably longer than a winter one. This was not considered a defect, the hours described the structure of the day rather than an abstract quantity of time. The day had a shape. Hours mapped that shape.

Equal hours were a Babylonian contribution. Working with their base-60 mathematical system, Babylonian astronomers divided the full day into 24 equal parts regardless of season. This required mechanical timekeeping. You cannot read an equal hour from shadows or rising stars because an equal hour is an abstraction, a unit of uniform duration that nature does not directly display. When water clocks became reliable enough to measure equal intervals, the Babylonian equal hour became the standard.

The hours, minutes and seconds we use today all carry Babylonian structure. There are 60 minutes in an hour and 60 seconds in a minute because Babylonian mathematics used base 60. The same reasoning gives us 360 degrees in a circle. The sky and the clock are divided by the same ancient arithmetic, inherited from astronomers pressing observations into clay beside the Euphrates.

Did You Know? The oldest surviving star clocks were painted on the inside of Egyptian coffin lids, dating from around 2150 to 1800 BCE. The decan star charts were placed there so the dead could tell the time in the afterlife. The same twelve stars that divided the night for the living would divide it for the dead. The system worked so well it was later painted on tomb ceilings, temple walls and water clocks. The 24-hour structure of your day was first recorded inside a coffin.

14 The World Adopts a Calendar

A few evenings later. Dad has an old world map spread across the table and is marking dates in different countries in pencil: 1582, 1700, 1752, 1873, 1918, 1949. Arjun traces the dates with his finger.

Arjun: When did the whole world agree on the same calendar?

Dad: It hasn't entirely. But most countries use the Gregorian calendar for civil purposes now.

Arjun: Why did some countries wait so long?

Dad: Politics, mostly. Protestant countries refused to adopt a Catholic reform regardless of its astronomical merits. Japan adopted it in 1873 as part of the Meiji government's modernisation programme. Russia adopted it in 1918, right after the Revolution.

Navya: Did they throw away their old calendars?

Dad: No. Almost every country that adopted the Gregorian calendar for civil use kept its traditional calendar for festivals. Japan still uses the lunisolar calendar for some regional festivals. China still uses the Chinese lunisolar calendar. The Gregorian calendar is for the government office and the train timetable. The traditional calendar is for the festival and the wedding.

Arjun: So most people in the world live by two calendars at once.

Dad: Most people in the world. Yes.

The global adoption of the Gregorian calendar took nearly four centuries and followed a pattern shaped more by political and religious alignment than by astronomical conviction.

Catholic countries in western and central Europe adopted the reform immediately in October 1582, Spain, Portugal, the Italian states and the Polish-Lithuanian Commonwealth. France followed in December. Protestant territories held out, viewing the reform as a Catholic imposition emanating from Rome. The German Protestant states adopted the Gregorian calendar only in 1700, when the accumulated Julian drift had grown to eleven days and was causing serious problems for commerce and legal correspondence with Catholic neighbours. Even then they initially refused to adopt the Catholic method for calculating Easter, producing a modified calendar they called the Improved Reckoning that disagreed with Catholic Easter in some years.

Britain's adoption in 1752 required an Act of Parliament. The Calendar New Style Act removed eleven days from September 1752, so that September 2nd was followed directly by September 14th, and shifted New Year's Day from March 25th to January 1st. Japan adopted the Gregorian calendar in January 1873. Russia adopted it in February 1918. The new Soviet government changed it partly because the continued use of the Julian calendar, by then thirteen days behind, signalled alignment with the Orthodox Church the Bolsheviks were actively suppressing.

In nearly every case, adoption of the Gregorian calendar for civil purposes was accompanied by the continuation of traditional calendars for religious and cultural life. This dual-calendar arrangement is now standard worldwide. The result is that most of the world's population navigates two calendar systems simultaneously, moving between them as context requires.

The world today runs two calendar systems simultaneously. Gregorian for civil administration and traditional for cultural and religious life.

One unexpected trace of the adoption survives today. Before France moved to the Gregorian calendar, the new year fell on April 1st. When January 1st became official, people who still celebrated in April were mocked as fools. One popular theory links April Fool's Day to this calendar confusion, though its true origin is uncertain.

True Story: George Washington was born on February 11th 1731 by the Julian calendar and February 22nd 1732 by the Gregorian. Two birthdays for one man. When Britain adopted the Gregorian calendar in 1752, Washington's birthday shifted forward eleven days. He acknowledged both dates but preferred February 22nd. Presidents' Day in the United States still marks that date.

15 Leap Years: The Full Rule

The following afternoon. Arjun is at the table with his notebook open, writing the leap year rule step by step in very neat handwriting. He always writes a rule out before he tests it. Dad is reading nearby and watching without appearing to.

Arjun: *(reading back)* Every four years, except century years, except centuries divisible by 400.

Dad: Correct. 1900 was not a leap year. 2000 was. 2100 will not be.

Arjun: Why did computers have a problem with this?

Dad: Much of the software written in the 1970s and 80s stored the year as two digits to save memory. When 2000 approached, these systems faced a question they hadn't been designed for: is "00" the year 1900 or 2000? It matters because 1900 was not a leap year but 2000 was. A system that confused the two would make errors in any date calculation involving February 29th.

Arjun: How much did it cost to fix?

Dad: Estimates range from 300 to 600 billion dollars globally.

Silence.

Navya: Calendars are expensive.

The Gregorian leap year rule has three layers. Every year divisible by 4 is a leap year. Except century years, which are not leap years. Except century years divisible by 400, which are. This produces 97 leap years in every 400-year period and an average year of 365.2425 days, within 26 seconds of the true tropical year. The Gregorian calendar will not need a comparable correction for approximately 3,000 years.

The century exception catches almost everyone out. Most people know the four-year rule and assume century years must be special leap years. The rule specifies the opposite. 1900 was not a leap year. 2100 will not be. 2000 was, because it is divisible by 400. The practical test, divide by 400 first. If yes, leap year. If by 100 but not 400, not a leap year. If by 4 but not 100, leap year. Everything else is not.

There are approximately 5 million people alive today born on February 29th. They are called leaplings. In most countries they are legally considered to age on February 28th in non-leap years, though some jurisdictions use March 1st. Gilbert and Sullivan's 1879 comic opera The Pirates of Penzance builds its entire plot on this ambiguity. A young man apprenticed to pirates until his 21st birthday discovers he was born on February 29th and is therefore technically only five years old. He cannot leave. The pirates have found the loophole in the calendar.

The leap day solves one problem but cannot solve everything. The Earth's rotation is gradually slowing due to tidal friction from the Moon. Atomic clocks, invented in the 1950s, are now so precise that they outrun the planet. Since 1972, scientists have periodically added a leap second to keep atomic time aligned with astronomical time. Twenty-seven leap seconds have been added so far. Unlike the leap day, which follows a fixed rule centuries in advance, leap seconds are announced only months ahead. The internet handles them poorly. Systems that expect 60 seconds in every minute occasionally crash when a minute contains 61. The leap second is being phased out by 2035.

The sky keeps moving. Every calendar correction is a negotiation with a universe that does not count in whole numbers.

Did You Know? February 30th has existed exactly once. In 1712 Sweden and Finland, having fallen behind in the transition from Julian to Gregorian calendar, added an extra day to February to help catch up. That year had a February 29th and a February 30th. No other year in recorded history has had a February 30th. It worked, Sweden rejoined the Gregorian calendar that year and has not needed a February 30th since.

16 Railways, Meridians and the Wristwatch

Dad is wearing an old mechanical wristwatch. He takes it off and puts it on the table. Arjun turns it over. There is an inscription on the back in his grandfather's handwriting.

Arjun: Did everyone always agree on what time it was?

Dad: Not at all. Before railways each city set its own local time by the Sun. Noon was when the Sun was highest locally. Bristol noon was ten minutes after London noon. Every town was slightly different.

Arjun: Which made train timetables impossible.

Dad: Dangerous. A head-on train collision in Rhode Island in 1853 killed fourteen people. The inquiry found that both conductors believed they were running on schedule, on different local times. Four minutes difference. That collision became one of the most cited arguments for standardised railway time.

Navya: *(holding the watch to her ear)* Why is it on your wrist? Why not in your pocket?

Dad: The First World War. Officers in the trenches needed to coordinate attacks to the minute without stopping. Wristwatches became standard military kit around 1915. After the war they became fashion items and replaced the pocket watch within twenty years.

Arjun: So standardised time and personal timekeeping both came from the same few decades.

Dad: The railways made precise time necessary. The First World War made it personal.

For most of human history, time was local and approximate. Noon occurred when the Sun was at its highest above the local horizon. In a world of slow travel the differences between towns, perhaps a few minutes at most between distant cities, were too small to matter. A farmer, a sailor, a merchant. None of them needed their clock to agree with anyone more than a day's walk away.

Railways destroyed this comfortable vagueness within a single generation. A train could cover in two hours what a horse took two days to walk. If a station in Bristol set its clock to Bristol time and a station in London set it to London time, the timetable connecting them was telling two truths simultaneously and neither fully. Railway companies initially published timetables in local times. Passengers missed trains. Connections failed. And on August 12th 1853, two trains collided head-on near Providence, Rhode Island, because the conductors were working from different local times four minutes apart. Fourteen people died. The inquiry's conclusion was direct, standardised time was not a convenience. It was a safety requirement.

The International Meridian Conference in Washington in 1884 formalised what railways had already made inevitable. A global system of 24 time zones measured from the Greenwich Meridian. France, which had used the Paris Meridian for centuries, abstained from the vote. China, which spans what would be five time zones, officially uses a single national time for political unity. Beijing time, always.

The wristwatch arrived as a military tool. Pocket watches required officers to stop and open them under fire, impossible during a coordinated attack. Specialist wristwatches were issued to military officers from around 1915. After the war they became civilian fashion items. Within two decades the pocket watch had essentially disappeared. The precision that industrial warfare demanded became the precision of daily life.

The precision of time zones and the coordination of a world that moves at the speed of trains, then planes, then data. None of these were planned. They were responses to problems that only became visible as travel and communication accelerated beyond the tolerance of approximate time. The calendar did not create modernity. Modernity demanded a more precise one.

True Story: On August 12th 1853 two trains collided head-on near Providence, Rhode Island, killing 14 people. The inquiry found both conductors believed they were on schedule, on different local times. The gap was four minutes. The collision became the strongest early argument in the United States for standardised railway time.

Story: The Imperial Bureau of Astronomy
China · Han Dynasty through Qing · 206 BCE to 1912 CE

The young astronomer has checked the eclipse calculation four times. He is going to check it once more.

His senior colleague watches him without comment. He understands. They all understand, in the Bureau of Astronomy. The stakes are not abstract. Three years ago, an astronomer in the eastern office predicted a solar eclipse for the fifteenth day of the third month. It occurred on the seventeenth. Two days of error. The man was demoted to the provinces. His family moved with him.

The young astronomer checks the numbers again. The arithmetic is clean. The eclipse will occur on the appointed day, in the appointed hour, in the appointed duration. He writes it out carefully, in the formal hand required for official documents. He rolls the scroll and hands it to his senior colleague, who will countersign it and pass it up the chain that eventually reaches the emperor.

An error is not merely a miscalculation. It is evidence that the emperor has lost alignment with the cosmic order, that Heaven no longer endorses the dynasty. This is what the calendar means in China. It is not an administrative convenience. It is a political document, renewed every year, that says. Things are as they should be.

This is why the Bureau's records are the most complete astronomical archive in the world. Every eclipse, every comet, every sunspot noted over two thousand years, recorded not because anyone was curious but because their lives depended on getting it right. Precision enforced by fear is still precision.

The young astronomer looks at the rolled scroll in his colleague's hand.

"It is correct," he says.

He almost believes it.

The Bureau's records survive. Modern astronomers working on the history of supernovae, comets and eclipses still consult them. The fear that produced them is gone. The precision remains. The young astronomer checking his calculations one more time contributed to a scientific archive that has outlasted the dynasty, the empire and the civilisation that demanded it.

ARC 5
THE CORE CONFLICT

17 The Calendar That Chose the Moon

A clear night. Mum has come outside and is sitting in the fourth chair with her tea. Oliver is still here. Arjun has made a table in his notebook. The Gregorian date of Eid al-Fitr for each of the last twelve years. The eleven-day drift is visible.

Oliver: I've been meaning to ask. Why does Ramadan keep moving? Every year it's in a different month.

Arjun: *(showing the notebook)* Eleven days per year. Look. Thirty-two and a half years for a full circuit.

Oliver: But the Indian and Chinese calendars correct for the drift. Why doesn't the Islamic calendar?

Dad: It was a deliberate choice. The Islamic calendar keeps festivals drifting through all seasons so that no generation ever has an easier Ramadan than any other. Fasting in summer in Melbourne might mean eighteen hours without water. In winter, maybe eleven hours in mild weather. Over a lifetime everyone experiences both.

Mum *(quietly)* It's a remarkable thing to decide. To choose difficulty on purpose so that no one has it easier than anyone else.

Oliver *(after a quiet moment)* That's actually kind of fair.

Navya nods from the grass.

The Islamic calendar follows the Moon alone. Twelve lunar months totalling 354 days per year, with no correction for the solar year. Islamic festivals therefore move backward through the Gregorian calendar by approximately eleven days annually. Over 32.5 years they complete a full circuit through every season and return to where they started.

This drift is visible in raw data. Eid al-Fitr fell in late August in 2011. By 2014 it had reached late July. By 2017 it was in late June. By 2023 it was in late April. The festival moves steadily backward, completing a full revolution around 2044. Anyone born today will see Ramadan in every season within their lifetime. The calendar is not drifting by accident. It is rotating by design.

The reasoning is often understood in terms of fairness. No community should permanently fast through short, cool winter days while another always fasts through long summer heat. The purely lunar calendar distributes the full range of conditions across every generation and every geography. Over time, this results in Ramadan being experienced in every season, a feature many scholars and observers interpret as promoting balance rather than favouring any particular time of year.

The Islamic calendar is also authenticated differently from any other major calendar. The beginning of each month is confirmed by the sighting of the crescent moon. In many communities two witnesses must observe it independently before the month is officially declared. The calendar cannot be confirmed from a table alone. It requires eyes. The sky must be consulted directly, and the month begins only when someone looks up and sees the crescent. A billion people begin and end Ramadan reading the same sky.

The communal act of looking is itself part of the calendar's meaning. The fast begins when the crescent appears. It ends when the next crescent appears. The calendar lives in the sky, not on the wall. No printed grid is required and no calculation is sufficient by itself. The month is declared only when the sky confirms it, and not before.

True Story: The Ottoman siege of Malta in 1565 occurred during Ramadan. Ottoman soldiers conducted assault operations while fasting through Mediterranean summer heat. Contemporary accounts from both sides document that the battle's rhythm was shaped by the fasting schedule. The pattern of engagements followed the pre-dawn meal, the sunset fast-breaking and the night prayers. The calendar had placed maximum religious obligation at maximum military difficulty.

18 Why Every Farmer Needed the Sun?

Dad has brought out seeds in small paper envelopes from the garden shed. He puts them on the table. Navya immediately picks one up and shakes it near her ear.

Dad: What do seeds need before they will germinate?

Arjun: The right season. Temperature, day length, soil warmth.

Dad: All solar signals. A seed cannot distinguish the fourth lunar month in winter from the fourth lunar month in summer.

Arjun: So any civilisation that farmed had to keep its calendar anchored to the solar seasons.

Dad: Every major farming civilisation that tried a purely lunar calendar either added solar corrections or eventually gave it up. The Babylonians, Indians, Chinese and Hebrews all arrived at lunisolar systems independently. The sky forced similar solutions.

Navya: The sky wins.

Dad: The sky always wins.

Agriculture is inherently solar. The signals that govern plant growth, soil temperature, day length, frost timing and monsoon arrival, are all driven by the Sun's position and Earth's axial tilt. A purely lunar calendar drifts against the solar year by eleven days annually. Within three years it has shifted by a full month. Within a generation it has rotated through every season. A farmer following a purely lunar calendar would eventually be planting winter crops in summer and harvesting in the wrong season entirely.

Every major agricultural civilisation on Earth independently developed a lunisolar calendar rather than a purely lunar one. The Babylonians had an early lunar calendar and by 500 BCE had developed a systematic intercalation cycle to keep it aligned with the agricultural seasons. Ancient India evolved the lunisolar panchang. Ancient China developed a lunisolar system incorporating the 24 solar terms, the jiéqì, which divide the solar year into agriculturally meaningful segments. Rain Water, Grain Buds, Great Heat, Frost's Descent, Major Snow. The Hebrew calendar developed its lunisolar structure to ensure Passover always falls in spring, when the barley harvest is ready.

The 24 solar terms are a particularly clear illustration of the solar year's priority even within a primarily lunar system. Each term marks a specific agricultural moment. These terms are still printed in Chinese calendars and used by farmers today. The lunar months carry the festivals. The solar terms carry the farming. The two systems run in parallel on the same calendar page.

The independence of these solutions matters. The Babylonian, Indian, Chinese and Hebrew traditions had no significant astronomical contact during the periods when they were developing their intercalation systems. They were responding to the same sky and arriving at the same structural conclusion. Twelve lunar months fall eleven days short of one solar year, and any calendar built around farming must correct that gap. The sky forced similar answers from every civilisation that planted crops beneath it and watched carefully enough.

The tension resolved differently in different cultures. Some honoured both the Moon and the Sun through lunisolar systems. The Islamic tradition honoured the Moon alone. The Gregorian tradition set the Moon aside entirely. All three choices remain in active use. Agriculture, where it continues, still answers to the Sun.

Around the World: China's 24 solar terms were inscribed on UNESCO's list of Intangible Cultural Heritage in 2016. They have been used continuously for over 2,000 years. The lunar months carry the festivals. The solar terms carry the farming. Both still appear on the same printed calendar page.

19 Easter: The Date That Needs Three Calendars

Dad has drawn three overlapping circles on the back of the envelope. He labels them: Solar Year, Lunar Cycle, Seven-Day Week. Where all three overlap he draws a small star and slides it across to Arjun.

Arjun: What is that?

Dad: Easter.

Arjun: Easter needs all three?

Dad: First Sunday after the first full moon after March 21st. One rule. Three calendars.

Arjun works through it.

Arjun: March 21st is the spring equinox: the solar year. The full moon is the lunar cycle. Sunday is the seven-day week. Easter cannot be fixed because all three have to line up simultaneously.

Dad: And they only line up in the same place every nineteen years.

Navya: *(without looking up)* That's why everyone argues about Easter.

Dad: That is exactly why.

Easter is the most computationally complex date in the Western calendar. Its calculation requires the solar year, the lunar cycle and the seven-day week to be satisfied simultaneously. The rule is simple to state, Easter falls on the first Sunday after the first full moon after March 21st. The consequences of that rule are not simple at all.

March 21st is the fixed ecclesiastical spring equinox, established by the Council of Nicaea in 325 CE when it ruled that Easter should be separated from Passover and anchored to the spring equinox instead. The full moon in the rule is not the astronomical full moon but an ecclesiastical full moon, calculated using the Metonic cycle from Idea 24. The Sunday is the seven-day week. Three independent cycles, none of which divides evenly into the others, must align before Easter can be placed.

The result is that Easter can fall anywhere between March 22nd and April 25th, a range of 35 days. No other major Western festival has a comparable range. Christmas is always December 25th. Diwali stays within a two-month window. Easter wanders across five weeks because three independent cycles pull in different directions and only occasionally agree.

The separation from Passover at Nicaea in 325 CE created a further problem, different Christian communities used different tables to calculate the ecclesiastical full moon and therefore arrived at different Sundays. By 664 CE the disagreement had reached a point where King Oswiu of Northumbria and Queen Eanflaed were celebrating Easter on different Sundays simultaneously in the same kingdom. The Synod of Whitby was called to resolve it. The king ruled in favour of the Roman calculation. The Celtic church submitted. Eastern and Western Christianity still calculate Easter by different methods, which is why Eastern Orthodox Easter often falls on a different Sunday from Western Easter. The three calendars were reconciled in 664 CE for the Western church. They were not quite fully reconciled. Eastern and Western Christianity remain on different Easter calculations today, and in some years the difference is a full month.

True Story: In 664 CE, King Oswiu of Northumbria was feasting to celebrate Easter while Queen Eanflaed was still fasting through Palm Sunday. The same festival, the same kingdom, two different Sundays. The Synod of Whitby was convened specifically to resolve this. Oswiu ruled in favour of Rome. The Venerable Bede records it as a decisive moment in the early English church.

20 Solar, Lunar, Lunisolar: The Full Comparison

Late afternoon, quieter than usual. Dad has produced a small notebook from his pocket and drawn a simple three-row grid in it - solar, lunar, lunisolar. He slides it across to Arjun. Navya clocks the notebook immediately.

Navya: *(to Arjun, quietly)* He prepared.

Arjun: Dad. We're in the garden.

Dad: *(opening the notebook)* I have been thinking about how to explain this for some time.

Navya: *(to Arjun)* He really prepared.

Arjun: *(finally)* Can we just do a clean comparison? Solar, lunar, lunisolar: what does each do and what does it cost?

Dad: *(slight smile)* That is exactly the question I have been waiting for.

Arjun studies the grid. After a moment he looks up.

Arjun: You have three rows. Day, month, year. But what about the week?

Dad: Because no calendar system produced it. The day comes from Earth's rotation. The month from the Moon. The year from the Sun. The week comes from none of these. It has no astronomical basis at all.

A solar calendar tracks the Earth's orbit around the Sun. Its months are administrative divisions with no astronomical basis. The year corresponds precisely to the solar year which means the calendar stays permanently aligned with the seasons. The price, a complete disconnect from the Moon. A solar calendar user who wants to know the Moon's phase must consult a separate source.

A lunar calendar tracks the Moon's phases. Its months are real astronomical events beginning at the new moon. The year of twelve such months is 354 days, eleven days shorter than the solar year. The calendar drifts through the seasons at eleven days per year, completing a full circuit in approximately 32.5 years. Festivals stay anchored to the lunar cycle but move freely through the solar seasons.

A lunisolar calendar tracks both. Its months are real lunar months. But it inserts an extra month every two to three years to bring the accumulated drift back into alignment with the solar year. Over any extended period the lunisolar calendar averages a year length close to the solar year, keeping festivals in their seasonal windows. The price is considerable computational complexity.

Type	Year	Months	Seasons	Example
Solar	365.2422 days	Administrative	Fixed	Gregorian
Lunar	354 days	Real lunar	Drifts	Islamic
Lunisolar	365.2422 avg	Real lunar	Corrected	Indian, Chinese, Hebrew

No type is objectively superior. Each makes a different trade-off. The Gregorian calendar achieves maximum administrative simplicity and permanent seasonal stability at the cost of any connection to the Moon. The Islamic calendar achieves perfect lunar accuracy and deliberate seasonal drift. Lunisolar calendars achieve both lunar accuracy and seasonal stability at the cost of computational complexity. The choice between them is a question about what a calendar is for.

The comparison also reveals what a calendar is for. A solar calendar maximises predictability. A lunar calendar maximises connection to a visible astronomical cycle. A lunisolar calendar attempts both, at the cost of greater complexity. Different cultures weighted these priorities differently. Reading a culture's calendar is reading its values.

There is, however, one unit of time that none of these three systems accounts for. The seven-day week. Unlike the day, the month and the year, it has no astronomical origin. How it came to govern the rhythm of every society on Earth is the subject of the following page.

Most of the world's festival calendar was designed in the northern hemisphere, where approximately 88% of the world's population has historically lived. The seasonal meanings embedded in the major festivals reflect northern hemisphere ecology. Diwali is an autumn harvest festival. It falls after the monsoon ends, when the harvest is in and the darkest night of the lunar month is lit with lamps. Holi is a late winter festival celebrating the end of cold. Chinese New Year is a late winter gathering as the agricultural year is quiet and spring is beginning.

When any of these festivals is observed in the southern hemisphere the seasonal meaning is completely inverted. Diwali in Melbourne falls in spring: new growth and blossom, not harvest and completion. Chinese New Year in Sydney falls in late summer, not late winter. The astronomical calculation of the festival is identical regardless of location, the new moon of Kartik is a global event. The Moon's phases are the same from Melbourne as from Mumbai. The seasonal context is entirely different.

Different diaspora communities have navigated this inversion in different ways. Some maintain the festival exactly as the original calendar prescribes. Others have begun incorporating local seasonal cues. Jacaranda blossom for Diwali in Melbourne, the smell of summer rain rather than autumn harvest. Neither response is wrong. Both are negotiations between an inherited calendar and the actual sky overhead. The same negotiation that produced the original calendar systems in the first place.

The southern hemisphere inversion is not a problem to be solved. It is a fact about where festivals were born and where they have since travelled. The communities that celebrate summer Christmas in Durban or spring Diwali in Buenos Aires are not celebrating incorrectly. They are doing what all diaspora communities have always done, carrying inherited time into a different sky and negotiating the gap between the calendar's memory and the season outside the window.

Did You Know? The city of Durban in South Africa has one of the largest Indian diaspora communities outside India. Diwali in Durban falls in late October or early November, the height of southern hemisphere spring. Local celebrations have developed their own seasonal imagery over generations, incorporating indigenous spring flowers alongside traditional diyas and fireworks. The festival and the southern sky have begun to find each other.

The Seven-Day Week: A Table With No Astronomical Excuse

The seven-day week is the only major unit of time with no basis in astronomy. The day comes from Earth's rotation. The month from the Moon. The year from Earth's orbit. The week comes from none of these. It is a human invention, and one that has proved utterly immovable.

Day	Named after	Planet	Norse / Germanic name
Sunday	The Sun	Sol	Sunnandæg
Monday	The Moon	Luna	Monandæg
Tuesday	Tiw (Norse god of war)	Mars	Tiwesdæg
Wednesday	Woden (chief Norse god)	Mercury	Wōdnesdæg
Thursday	Thor (Norse god of thunder)	Jupiter	Þūnresdæg
Friday	Frigg / Freya (Norse goddess)	Venus	Frīgedæg
Saturday	Saturn	Saturn	Sæternesdæg

The mechanism behind the weekday order is called the hora system, found in both Vedic and Babylonian astronomical traditions. A hora is a planetary hour. Each of the day's 24 hours is assigned to one of the seven classical planets in sequence, cycling continuously. The planet ruling the first hora of each day gives that day its name. Because 24 hours across 7 planets leaves a remainder of 3, the first hora of the next day falls three steps further along the sequence. This arithmetic produces the weekday order we still use. The Surya Siddhanta describes the system in Chapter XII. The lords of the days are reckoned in order fourth from Saturn downwards, and the lords of the hours commencing from Saturn downwards. Work through the numbers and Saturday, Sunday, Monday, Tuesday, Wednesday, Thursday, Friday emerge in exactly that order.

The Romans spread the system across their empire. As Christianity moved north into Germanic Europe, the Roman planet names were replaced with Norse equivalents. Mars became Tiw, Mercury became Woden, Jupiter became Thor, Venus became Frigg. Saturday kept Saturn because the Norse had no equivalent powerful enough to displace him.

The French Revolutionary Calendar of 1793 replaced the week with a ten-day décade. It lasted twelve years. The Soviet Union tried a five-day week in 1929 and a six-day week in 1931. Both failed within years. No calendar reform has ever successfully dislodged the seven-day week. It has no astronomical justification and has outlasted every rational attempt to replace it.

21 What Makes a Calendar Right?

A quieter evening. Dad is looking at the sky. Arjun is writing. Navya is building a small pile of stones from the garden path.

Arjun: Is there a right answer? Is one calendar more correct than the others?

Dad is quiet for a moment.

Dad: Correct by what measure? The Gregorian calendar is the most useful for global coordination. The panchang is the most astronomically detailed. The Islamic calendar is the most theologically consistent. The Chinese calendar is the most ecologically embedded. Which is correct depends on what you want a calendar to do.

Arjun: So the answer is no.

Dad: The answer is that the question is poorly formed. A calendar is a tool. Asking which calendar is correct is like asking which is the correct tool: a hammer or a scalpel.

Navya adds one more stone to her pile.

Navya: I want a calendar that gives me more weekend.

Dad: Several have tried. All have failed.

No calendar is astronomically perfect. The tropical year is 365.2422 days, a number with no exact representation in any system of whole months and whole days. Every calendar involves rounding and every approximation introduces drift. The Gregorian calendar is accurate to about 26 seconds per year. The Hebrew calendar's 19-year cycle produces an average year of 365.2468 days, slightly longer than the tropical year but stable enough for practical purposes. The Indian panchang tracks multiple cycles simultaneously with extraordinary precision but even it operates within the fundamental constraint that 12.3683 cannot be made whole.

Whether a calendar is correct depends entirely on what it is asked to do. For global civil coordination the Gregorian calendar is indispensable. Its near-universal civil adoption makes it the only workable standard for international business, aviation and diplomacy. For astronomical transparency the Indian panchang has a strong claim, it does not simplify or hide and it treats the reader as someone who can handle the real complexity of the astronomical situation. For theological consistency both the Islamic calendar, purely lunar for fourteen centuries, and the Hebrew calendar, lunisolar with Metonic intercalation for sixteen centuries, have maintained their founding principles without significant modification.

The diversity of answers is not a failure to reach consensus. It is a record of different civilisations asking genuinely different questions. What should the year be anchored to? What should a month represent? What obligations does a calendar impose and on whom? These are not astronomical questions. They are questions about values.

The question of what makes a calendar right has no single answer. The Gregorian calendar is correct for global coordination. The panchang is correct for astronomical transparency. The Islamic calendar is correct for theological consistency. Each has been maintained for centuries by hundreds of millions of people. All of them describe the same sky. None of them agrees with the others about what day it is. A calendar is not a fact about the universe. It is a decision about what to count.

Strange But True: In 1849 the French philosopher Auguste Comte designed a Positivist Calendar. 13 months of exactly 28 days, each month and day renamed after a great historical figure. Moses, Homer, Aristotle, Archimedes and Shakespeare each got a month. Sundays were replaced with days named after scientists and philosophers. The calendar was mathematically clean, culturally ambitious and adopted by exactly one person, Comte himself. He used it until his death in 1857. Every calendar that has tried to replace human habit with pure logic has failed. Comte's lasted eight years and required the inventor.

Story: Al-Biruni's Journey to India
Persia and India · 1017 to 1030 CE

The Sanskrit is coming slowly. Al-Biruni traces the characters with one finger, his lips moving. Across the table, the Brahmin mathematician watches him with an expression that is not quite patience and not quite impatience. The face of a teacher who is not sure the student is worth the effort.

"Again," al-Biruni says. "The correction for the leap month. Show me the calculation from the beginning."

The Brahmin sighs faintly. He has explained this three times. But he picks up his stylus and begins again, drawing out the steps in the dust of the tabletop. Al-Biruni watches with the focused attention that has, over the past weeks, begun to change how the Brahmin sees this visitor from the west. He does not argue. He does not insist his own method is better. When the Indian calculation gives a result different from the Islamic one, he writes down both and asks questions. When the Indian method is more precise, which it often is, he says so plainly. He seems to find this delightful rather than threatening.

Al-Biruni has been in India for three years. He has learned enough Sanskrit to read astronomical texts without a translator. He has visited scholars in several cities. He has compared the Indian values for the length of the year, the precession of the equinoxes and the Moon's orbital period against Greek and Islamic values, and found them arrived at independently, through different methods, often agreeing to within minutes.

The Brahmin finishes the calculation and looks up.

"You understand now?"

Al-Biruni is still writing. "I am beginning to," he says. "I think it will take a few more years."

The Brahmin looks at him for a moment. Then, for the first time, he almost smiles.

His book, the Indica, is still read. It remains one of the most careful and respectful accounts of Indian science ever written by an outsider. Al-Biruni reached the same conclusion this book is trying to reach. That different civilisations looking at the same sky arrived at different answers, and that understanding those answers requires curiosity rather than judgement. He just had to travel further to make the point.

ARC 6
THE SOLUTION

22 The Month That Repeats

An afternoon play date. Mei is over. Navya and Mei are at the table doing homework. Arjun and Dad are outside. Mei looks up from her worksheet.

Mei: My mum said this year has an extra month in the Chinese calendar. Like, there's a repeated month.

Arjun: An Adhik Maas? We have that in the Indian calendar too.

Mei: Same idea? Mum just said the year is longer this year.

Dad: Yes, same idea entirely. The Indian calendar calls it Adhik Maas. The Chinese calendar calls it Rùn Yuè. Both are a thirteenth month inserted when the lunar calendar has drifted too far from the solar year.

Mei: Why does it have the same name as the month before it?

Dad: Because it doesn't have its own name. It borrows the name of the month it follows. It's the extra one, the repeated one.

Navya: *(looking up from her homework)* A bonus month.

Mei: *(to Navya)* Mum said we get extra time for things. Like a gift.

Dad looks at both of them.

Dad: In the Indian tradition it's considered sacred time. The extra month is called Purushottam Maas, the month of the Supreme Being.

Mei: Ours is just called the intercalary month. Not as good a name.

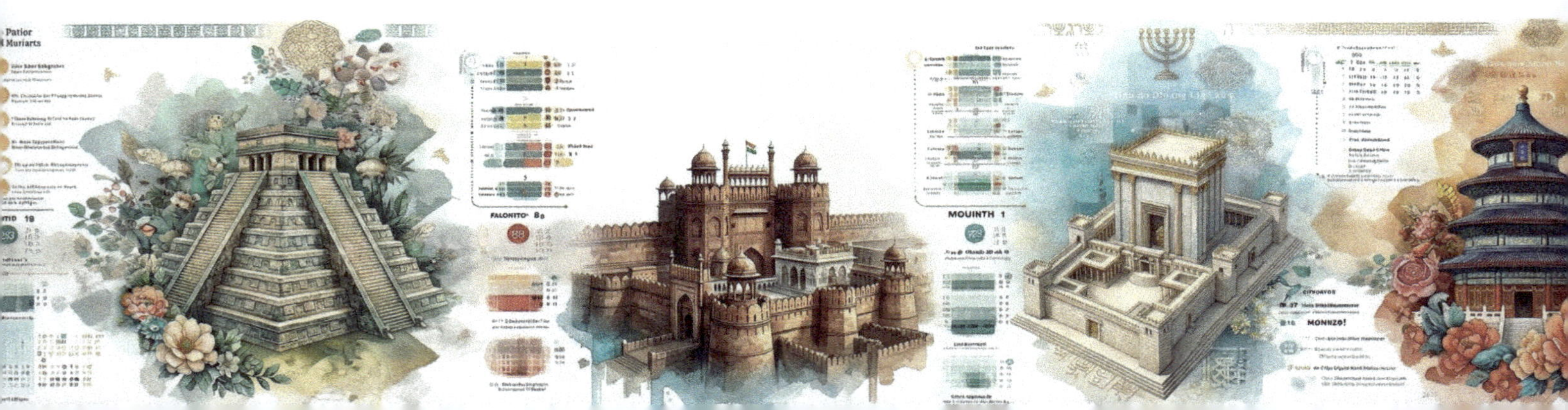

The leap month, an extra lunar month inserted periodically to bring the lunisolar calendar back into alignment with the solar year, was discovered independently by at least three major astronomical traditions. Indian, Chinese and Hebrew. Babylon preceded all three. Each arrived at the same structural solution through a completely different observational path.

In the Indian lunisolar calendar the extra month is called Adhik Maas, meaning additional month. The rule is astronomically precise. An Adhik Maas occurs whenever a lunar month passes without the Sun entering a new sign of the zodiac. Normally each lunar month corresponds to the Sun moving from one zodiacal sign to the next, a transition called a Sankranti. When a lunar month is too short to contain a Sankranti it is declared the extra month, the following month's name is reused and the drift is reset. This mechanism inserts a leap month approximately every 32.5 months, seven times in every 19 years.

In the Chinese lunisolar calendar the extra month is Rùn Yuè, the intercalary month. The Chinese system uses the 24 solar terms as its anchor. A lunar month that contains no principal solar term is designated the intercalary month, taking the name of the preceding month. The Chinese New Year's guaranteed arrival between January 21st and February 20th, year after year, is a direct consequence of this mechanism operating continuously in the background.

In the Hebrew lunisolar calendar the extra month is Adar Sheni, Second Adar. The Hebrew calendar adds Adar Sheni in seven of every nineteen years according to a fixed table derived from the Metonic cycle. The rule ensures that Passover always falls in spring: if the standard calendar would place Passover before the vernal equinox, Adar Sheni is inserted to push it after.

The independent discovery of the leap month by multiple civilisations is parallel invention in its clearest form. The same problem, the same constraint, the same solution, found separately with no contact. The Babylonians, Indian astronomers and Chinese calendar-makers all arrived at the intercalated month because the sky left them no alternative. The solutions converge because the problem is universal.

Around the World: The year with an Adhik Maas is considered especially sacred in the Hindu tradition. The extra month is called Purushottam Maas, the month of the Supreme Being. Religious observances performed during this month are believed to carry special merit. The astronomical correction for a calendar problem became, in the Indian tradition, an additional gift of sacred time.

23 The Calendar as Resistance

Oliver and Mei are both over. Mei is showing Navya something on a piece of paper. Oliver is watching.

Oliver: *(to Dad)* My grandparents came from Ireland. They still celebrate some festivals on the old Julian dates, not the Gregorian ones. My mum says they never really accepted the calendar change.

Dad: That is more common than people think. The switch to the Gregorian calendar was not just a technical adjustment. For many communities it felt like an imposition.

Mei: My grandparents use the Chinese calendar for everything that matters. Gregorian is for work.

Arjun: Same with us. Two calendars running at once.

Oliver: Has it always been like that?

Dad: Since the colonial period, mostly. Wherever the British or French or Spanish administered, the Gregorian calendar followed. Not as a choice. As a condition of legal existence.

Oliver: And the old calendars?

Dad: Survived. In exactly the way Mei's grandparents use theirs. In the home. In the festivals. In the life the government could not reach.

Navya: *(without looking up)* The sky didn't change.

Dad: No. The sky didn't change.

Every calendar encodes a set of values. When one culture imposes its calendar on another, it is not merely changing the dates. It is asserting that one way of understanding time is the correct one. The history of calendar adoption is therefore also a history of power and of resistance.

The most systematic example is the colonial period. As European administrations extended across Asia, Africa and the Americas, the Gregorian calendar followed as an instrument of governance. Courts, land records, tax systems and contracts were all anchored to Gregorian dates. To function in the colonial system, colonised peoples had to operate in Gregorian time. Their traditional calendars were not banned in most cases. They were made irrelevant to power.

The response was consistent across cultures. The traditional calendar survived in private life. The Indian panchang continued to govern festivals, marriages and planting seasons through two centuries of British rule. The Chinese lunisolar calendar was never abandoned despite China adopting the Gregorian calendar for civil use in 1912. The Hebrew calendar has been maintained continuously through sixteen centuries of diaspora, surviving expulsions and persecution across dozens of countries. In each case the traditional calendar became an assertion of identity. A way of saying that official time and lived time are not the same thing.

The most complete example of calendar survival under suppression may be the Aboriginal seasonal calendars. These are not written systems. They encode knowledge of when animals breed, when plants fruit and when to fish, carried in song, story and ceremony. Two centuries of colonial suppression could not destroy them because they were not stored anywhere that could be confiscated. Communities forcibly relocated and barred from their country still passed this knowledge between generations. The sky they describe is the same sky it has always been.

Every failed calendar reform teaches the same lesson. A calendar is not an administrative tool that can be swapped by decree. It is a shared agreement about what time means. When to rest, when to celebrate, when to plant. Communities do not abandon that agreement easily. The sky does not change when the government changes the calendar. The festivals continue to fall where the sky places them.

Worth Knowing: Singapore officially uses the Gregorian calendar and additionally recognises Chinese New Year, Deepavali, Eid al-Fitr and Hari Raya Haji as public holidays, each calculated by a different calendar system. A 1968 parliamentary debate about public holidays became, in effect, a debate about which calendar traditions the state would officially recognise. Singapore kept all four, explicitly acknowledging that four independent calendar systems were operating simultaneously in one small city-state.

24 The Nineteen-Year Secret: The Metonic Cycle

Dad has brought out a calendar for the year nineteen years in the future and placed it beside this year's. The full moons fall on almost exactly the same dates. Arjun looks between them.

Arjun: Is there a pattern to when the leap months get inserted? A cycle?

Dad: Yes. And it's one of the most beautiful coincidences in the relationship between the Earth and the Moon. Nineteen solar years are almost exactly equal to 235 lunar months. After nineteen years the Moon's phases fall on almost precisely the same calendar dates as they did nineteen years earlier.

Arjun makes the connection immediately.

Arjun: So if you know the leap month pattern for one nineteen-year cycle you know it for every cycle after that. Indefinitely.

Dad: The Hebrew calendar has used a fixed table built on this cycle since approximately the fourth century CE.

Arjun: Who discovered it?

Dad: A Greek astronomer named Meton announced it in Athens in 432 BCE. But Babylonian astronomers were already using it a century before he was born. Indian and Chinese astronomers knew it independently. It's something the sky eventually shows to everyone who watches carefully enough.

Navya, from the grass, without moving.

Navya: Nineteen years.

Nobody asks what she means.

The Metonic cycle is one of the most practically useful coincidences in observational astronomy. 19 solar years are almost exactly equal to 235 lunar months. The difference is approximately two hours over the full 19-year period, meaning that after 19 years the phases of the Moon recur on almost precisely the same dates of the solar calendar as they did at the start.

For lunisolar calendar design the implications are profound. If 19 solar years equal 235 lunar months, and a regular year has 12 lunar months, then 19 years requires 235 minus 228 equals 7 additional months. Seven leap months in every 19-year cycle. The Hebrew calendar uses exactly this structure, seven leap months in years 3, 6, 8, 11, 14, 17 and 19 of each 19-year cycle. The pattern has been used continuously since approximately the fourth century CE. The Hebrew calendar's average year of 365.2468 days differs from the tropical year by only about 6 minutes annually.

The Greek astronomer Meton of Athens announced the 19-year cycle in 432 BCE at the Olympic Games. Babylonian astronomical records show that Babylonian astronomers were already applying a 19-year intercalation cycle by at least the fifth century BCE. Indian astronomical texts record related cycles. Chinese records show awareness of the 19-year relationship independently of both Babylonian and Indian traditions.

The convergence of Babylonian, Greek, Hebrew, Indian and Chinese astronomy on the same numerical relationship is one of the most striking examples of independent discovery in the history of science. The sky posed the same question to every careful observer. The same number emerged as the answer every time.

The Metonic cycle's discovery was so practically useful that it was independently found at least three times: by Meton of Athens around 432 BCE, by Babylonian astronomers working from somewhat earlier records, and by Indian astronomers who encoded it in the structure of the panchang. The Hebrew calendar, which has used a fixed Metonic table since approximately 359 CE, can calculate the date of Rosh Hashanah for any year in the next several thousand years without a single additional astronomical observation. The cycle runs itself.

The Number: 19 years = 235 lunar months, with an error of only two hours. Meton of Athens announced this cycle publicly at the Olympic Games of 432 BCE. He may have chosen the venue deliberately, the Olympic Games were themselves scheduled using the lunar calendar, and the audience included astronomers from across the Greek world. The same cycle he announced that day still governs the dates of Rosh Hashanah and Passover today. A calculation made in Athens in 432 BCE is still running.

25 The Sky Was Never Simple

Everything is on the table but nothing is open. The panchang is closed. The globe is still. The star chart is folded. Mum is outside in the fourth chair. The light has the quality of something ending.

Arjun: Every calendar system is trying to answer the same question. They've all just decided that different things matter most.

Dad: Yes. The Gregorian calendar decided to make civil coordination as simple as possible. The panchang decided to show the astronomical truth, all of it. The Islamic calendar decided to make religious obligation equal across all times and places. The Chinese calendar decided to preserve both the Moon and the agricultural seasons. The Hebrew calendar decided to keep the festivals in their ecological seasons while preserving the lunar cycle.

Pause.

Arjun: They're all right.

Dad: *(quietly)* They're all right. And they all have the same moon above them.

Navya *(on her back)*: I thought the moon was rude.

Dad laughs, the same laugh as the first evening. Then Mum. Then Arjun. The panchang sits closed. The same crescent moon that was in the sky on the first evening is there again.

We began this book with a question that belongs to every culture on Earth. Why do festival dates keep changing? The child asking it might be in Mumbai or Melbourne, in Jakarta or Jerusalem, in Lagos or London. The question is the same. And we have now seen that it contained a false assumption. Festival dates do not change randomly. They follow precise astronomical rules that any careful observer can learn and predict. Diwali is exactly where it always is, on the new moon of Kartik. Chinese New Year is exactly where it always is, on the second new moon after the winter solstice. Ramadan begins exactly where it always does, when the crescent moon appears after the new moon of the ninth Islamic month. Easter falls on the first Sunday after the first full moon after Spring Equinox. Every time.

What shifts is the vantage point. The Gregorian calendar follows a different cycle from these festivals. It tracks the Sun, treats months as administrative units and starts its year on a date chosen by Roman consuls two thousand years ago. When a lunar or lunisolar festival is viewed through this calendar it appears to move. In reality the Gregorian calendar is moving past a festival that is standing still in its own astronomical frame.

The deeper realisation is about the sky itself. The sky is genuinely complex. A year is not a whole number of months. A month is not a whole number of days. The three fundamental cycles are incommensurable. No whole-number connects them and none ever will. Every calendar is an approximation and it involves a choice about what to preserve and what to sacrifice.

Every culture that looked up at the same sky and tried to count what it saw arrived at a different answer. And every answer was sincere. The Moon waxes and wanes on the same 29.53-day cycle from every place on Earth. The answers differed because the questions differed. Because what people needed from time, and what they valued about the sky, was shaped by who they were, where they lived and what they believed. The calendars are not competing mistakes. They are portraits.

The sky above you tonight is the same sky that Egyptian priests watched for the rising of Sirius, that Aboriginal Australians read for the running posture of an emu made of darkness, that Babylonian astronomers mapped onto clay tablets and that every family in every culture has ever looked up at and wondered about. The calendars are different maps of the same territory. The territory does not change. Only the maps do.

> **To Think About:** Look up tonight. The Moon is somewhere in its cycle. If you know roughly which night of the lunar month it is, you are reading the same sky that every calendar on Earth was built to describe.

Conclusion

The same backyard. Arjun, Navya and Dad are sitting in the backyard looking at the sky and waiting for the constellation of Southern Cross to appear. Mum has also joined in. Nobody is talking. The silence is comfortable.

Arjun: *(after a while)* If I wanted to tell someone what year it is, what year is it actually?

Dad does not answer immediately.

Dad: In the Gregorian calendar, this year. In the Hebrew calendar, a different number. In the Islamic calendar, another number entirely. In the Indian traditional calendar, yet another. All of them are right.

Arjun: That's not one answer.

Dad: No. It's four answers. That might be more honest than one.

The Southern Cross is clearly visible now.

Navya: *(without moving)* I know what year it is.

Nobody responds. She does not rush.

Navya: It's the year of the stars.

A long silence.

Dad: *(smiling)* That might be the most accurate answer of all.

Mum looks at Navya. Then at the stars. She does not say anything.

We are left at the end not with a single correct calendar but with a deeper understanding of why no single calendar can be correct. The sky is exactly as complicated as it is. It offers three incommensurable cycles, day, month and year, and no arrangement of whole numbers can perfectly describe all three simultaneously. Every calendar is a negotiation with this fact. Every calendar is a portrait of its makers.

The Indian panchang shows everything. The lunar phase, the Moon's nakshatra and the Sun's position among the stars. Its complexity is its honesty. The Chinese lunisolar calendar preserves both the Moon and the agricultural seasons through an elegance of mechanism arrived at over centuries of careful observation. The Hebrew calendar has maintained its Metonic cycle with remarkable consistency for sixteen centuries. The Islamic calendar has maintained its purely lunar character for fourteen centuries, distributing the obligation of fasting equally across all seasons and all generations.

The Gregorian calendar, which most of us use for daily administrative purposes, is also a product of this long history. It has descended from the Egyptian 365-day year, reformed by Julius Caesar, corrected by Christopher Clavius and adopted across the world through European expansion and genuine administrative utility. It is convenient and internationally standardised. It is also quietly full of political accidents. Months named after Roman emperors, a new year on a date with no astronomical basis and a leap year rule that required a decade of argument to agree. It describes the sky about as accurately as a street map describes a city. Well enough to get around, not well enough to know where you actually are.

The Moon does not know about the Gregorian calendar or the panchang or the Metonic cycle. It rises when it rises, in the phase it is in, on its own ancient schedule. The sky was never consulted about our calendars. It was only observed. The observation is the same everywhere. The calendars are the record of what different people decided to do with what they saw.

The sky is honest. It is exactly as complicated as it is.
Calendars are what happens when humans decide to argue with that.

Rahul Agrawal is an Information Technology Consultant by profession and lives in Melbourne with his two children, Arjun and Navya. He has a deep interest in understanding the sky, time and the way ancient cultures made sense of the world through observation. Many of these ideas took shape through frequent conversations with his son Arjun, where simple questions about the sky, stars and the passage of time often led to thoughtful discussions and shared exploration.

Inspired by these interactions, Rahul developed a growing interest in calendars and Astronomy, especially how traditional knowledge connects the movement of celestial bodies with everyday life. He is particularly drawn to making these concepts simple and engaging for young minds, helping children understand how the sky can be read like a story. This approach aligns with the vision of the "One World" series, which introduces children to meaningful ideas from different perspectives in a clear and relatable way.

Through his work, Rahul aims to present these fascinating connections between the sky, time and human understanding in a way that feels accessible and enjoyable. By breaking down complex ideas into simple explanations, he encourages curiosity, observation and a sense of wonder about the world above us.

Also in the One World Series:

Australia: The Youngest Old Country explores 25 defining ideas that shaped the nation, from its ancient roots to its modern identity. Like this book, it dives into the conversations and ideas that have guided Australia's journey, revealing how the past continues to shape a country evolving under one shared sky.

If you enjoyed this book, please share it with friends and family and consider leaving a review, helping these ideas reach and inspire more young minds.